Innovations in Cell Research and Therapy

Edited by Zvi Loewy

Published in London, United Kingdom

IntechOpen

Supporting open minds since 2005

Innovations in Cell Research and Therapy
http://dx.doi.org/10.5772/intechopen.73444
Edited by Zvi Loewy

Contributors
Peter Quesenberry, Theo Borgovan, Lorin Crawford, Chibuikem Nwizu, Mónica Martínez Gallo, Marina García-Prat, Sanda M. Cretoiu, Maria-Simonetta Faussone-Pellegrini, Maria Giuliana Vannucchi, Lidia Ibba-Manneschi, Dragos Cretoiu, Mirko Manetti, Yihua Bei, Junjie Xiao, Sergio Oviedo, Raquel Guillamat-Prats, Antonio Artigas, Mitchell S. Cairo

First published in London, United Kingdom, 2020 by IntechOpen
IntechOpen is the global imprint of INTECHOPEN LIMITED, registered in England and Wales, registration number: 11086078, 7th floor, 10 Lower Thames Street, London, EC3R 6AF, United Kingdom
Printed in Croatia

British Library Cataloguing-in-Publication Data
A catalogue record for this book is available from the British Library

Additional hard and PDF copies can be obtained from orders@intechopen.com

Innovations in Cell Research and Therapy
Edited by Zvi Loewy
p. cm.
Print ISBN 978-1-78984-456-6
Online ISBN 978-1-78984-457-3
eBook (PDF) ISBN 978-1-83880-020-8

For EU product safety concerns:
IN TECH d.o.o., Prolaz Marije Krucifikse Kozulić 3, 51000 Rijeka, Croatia, info@intechopen.com or visit our website at intechopen.com.

Meet the editor

Dr. Zvi Loewy, a senior academic leader and an experienced global pharmaceutical–biotechnology executive, leverages a diversified background in big-pharma senior management, biotech startup creation, and academia. He is a true visionary, who has been the key inventor on many patents.

Dr. Loewy's international experience has included leading international research teams; championing the penetration and commercial launch of consumer healthcare products in China and developing markets; and leading open innovation in the Middle East.

Dr. Loewy is on the faculty of the Touro College of Pharmacy and New York Medical College; the boards of the Jerusalem College of Technology, the New Jersey Bioscience Incubator; and is an editor of the *Journal of Prosthodontics* and the journal *Microbiome*. He has published broadly and has more than 25 issued patents

Contents

Preface

In recent years, progress in cell biology has uncovered complex molecular networks that have elucidated the underpinning defects and therapeutic targets in biological systems. Advances in molecular biology and the pathology of disease have provided the foundation for the advent of therapeutic interventions and innovations designed to prevent or modify disease. Today, personalized medicine is delivering targeted therapies that are optimized for the patient's phenotype and genotype.

As our understanding of the broad field of "omics" manifested by genomics, transcriptomics, proteomics, and metabolomics has contributed significantly to twenty-first century therapeutics, so too has our understanding of fundamental cell biology. To that end, it became apparent that a need exists for capturing advances in cell structure, function, and therapy.

The dawn of cell therapies and medicines is captured in this new and exciting book entitled *Innovation in Cell Research and Therapy*. This book provides an in-depth description of some of the key issues and advancements in cell biology and its application to novel therapeutic approaches. Six chapters authored by experts in the field address the components of cells, cell structure, and the emergence of stem cells and their applications. The chapters provide state-of-the-art descriptions and summaries and identify the future needs in this continuously evolving field. This book is meant to raise awareness of the complexity associated with defining and characterizing the biological and pharmacological properties of cell therapy modalities.

Hardly a day passes without hearing of a biomedical advance. Innovation in the biosciences has progressed at a rapid speed. Given that, why should you read this book? What do we bring to the table? Our answer is that we are a unique team of global experts that come from diverse backgrounds. In this engaging book we deliver a comprehensive compendium of some of the most fundamental, important, and enabling advances in cell biology. As you read the book you will personally engage with leaders in the cell biology field. I am confident that your knowledge and insight into cell biology will be significantly impacted. Indeed, the dawn of cell therapies and medicines is captured in this new and exciting book.

I am humbled and most gratified to IntechOpen for inviting me to serve as editor of this work. My sincere thanks go to Josip Knapic, Author Service Manager, and Martina Usljebrka, Commissioning Editor, for all of their help and assistance throughout the entire process. I thank my colleagues at the Touro College of Pharmacy and New York Medical College for fostering an inspiring environment with a shared quest for biomedical advancements. Words cannot express my gratitude to my wife Brenda and our children and grandchildren for their constant support and encouragement. This book is dedicated to the memory of my wonderful parents Harry and Cecile Loewy. It is their value of education, and contribution to the field of education, that set me on my own personal path in academics.

Dr. Zvi Loewy
Professor of Pharmaceutical and Biomedical Sciences,
Associate Dean of Research,
Immediate Past Dean,
Touro College of Pharmacy,
New York, USA

Professor of Microbiology and Immunology,
New York Medical College,
New York, USA

Section 1

Cell Therapy

Chapter 1

Current Status of Stem Cell Therapy for Sepsis and Acute Respiratory Distress Syndrome

Raquel Guillamat-Prats and Antonio Artigas

Abstract

Sepsis and acute respiratory distress syndrome (ARDS) are life-threatening diseases with high mortality, around 40%, and morbidity in all the critical care units around the world. After decades of research, and numerous pre-clinical and clinical trials, sepsis and ARDS remain without a specific and effective pharmacotherapy and essentially the management remains supportive. Over the last years, cell therapies gained potential as a therapeutic treatment for ARDS and sepsis. Based on numerous pre-clinical studies, there is a growing evidence of the potential benefits of cell-based therapies for the treatment of sepsis and ARDS. Different cell subtypes have been used for the treatment of both syndromes; however, the major part of the studies is using mesenchymal stem/stromal cells (MSC). Also, other relevant groups performed some pre-clinical studies using induced pluripotent stem cells (iPSC) for the treatment of both syndromes and alveolar type II cells for ARDS treatment. Numerous questions need further study, including determining the best source for the progenitor cells isolation, their large-scale production, and cryopreservation. Also, the heterogeneity of patients with sepsis and ARDS is massive, and the stratification of the patients will help us to determine better the therapeutic effect of these cell therapies. In this review, we are going to describe briefly the different cell types, their potential sources, and characteristics and mechanism of action. We will review several pre-clinical and clinical studies in ARDS and sepsis.

Keywords: sepsis, ARDS, acute lung injury, cell therapy, critical care

1. Introduction

Sepsis and acute respiratory distress syndrome (ARDS) are life-threatening diseases with high mortality and morbidity in all the critical care units around the world [1–4]. Severe sepsis is a complex syndrome produced by the response to a systemic infection. The infection produces a general inflammatory response, such as tachycardia, elevated white cell count and systemic release of pro-inflammatory cytokines, and this can lead to an acute organ dysfunction. Sepsis is producing more than 5 million deaths per year worldwide [5–8].

The lung is one of the most affected organs during sepsis, and for that reason, one of the main indirect causes of ARDS is sepsis. ARDS can also be produced by a direct injury as a pulmonary infection or a trauma. ARDS is a multifactorial syndrome characterized by increased lung permeability, hypoxemia, the absence of cardiogenic pulmonary edema, the disruption of the alveolar-capillary barrier, and widespread inflammation. Every year between 1.5 and 4.5 cases per 100,000 inhabitants/year in Europe is detected and in the United States approximately 200,000 new cases are identified per year [1, 5, 8].

There is no definitive therapy that targets the underlying pathobiology of sepsis exists. Nowadays, the treatment is based on antibiotics, infection source control, fluid resuscitation, and organ support [9, 10]. Moreover, several patients die due to secondary infections during the year after the hospital discharge. After decades of research, and numerous pre-clinical and clinical trials, sepsis and ARDS remain without a specific and effective pharmacotherapy and essentially the management remains supportive. Nowadays, patients with ARDS have been treated with several ventilator interventions such as lower tidal volumes, higher positive end-expiratory pressure (PEEP), and adjuncts such as prone positioning, neuromuscular blockade, and extracorporeal membrane oxygenation [11–14].

Current advances in the study and knowledge of stem cells have permitted to start using them as a novel treatment for ARDS and sepsis. Based on numerous pre-clinical studies, there is a growing evidence of the potential benefits of cell-based therapies for the treatment of sepsis and ARDS. Several cell types are used in the last years for the treatment of both syndromes showing high efficiency [15, 16].

In animal models, cell therapies have demonstrated noteworthy therapeutic properties including the modulation of the immune system, the release of several factors with growth factor, and antibiotic and anti-inflammatory properties. Cell therapies were tested in several animal models, such as mouse, rat, sheep, and pig models, using several septic and ARDS models. Moreover, different cell types and different administration pathways (intravenously, intraperitoneal, or local administration into the lung) were used as a treatment.

Furthermore, in the last years, a couple of clinical studies started using cell therapies for the treatment of sepsis and ARDS, and some safety and efficient results are published.

This chapter summarizes the different progenitor cells that can be used as a therapy, the mechanisms of action, and the results in pre-clinical and clinical studies in ARDS and in sepsis and future directions.

2. Cell sources for transplantation

We define cell therapy as a therapeutic product containing cells, which usually is administered into the patients to replace or repair damaged tissues or cells. Nowadays, several diseases are treated with cell therapies, for example, bone marrow transplants for the treatment of some specific cancers.

In the last years, several researchers and physicians are working to convert some cell therapies from potential treatments to real therapies. There is an effort to detect the factors that the cells are secreting and have this beneficial effect and also to set up the safety and efficacy. Additionally, these therapies are really expensive, and there are several problems associated such as the difficulty to obtain, expand, purify, and manipulate these cells. So, we also have to work on the cost-effective options.

Cell therapies have shown their potential in biomedicine, and their utility for several indications has been demonstrated and this utility will expand in the future. Nevertheless, progressing cell therapies from bench to bedside takes decades of hard and slow work.

Several pharmaceutical companies hold a number of stem cell lines and work to advance in cell therapies. Finding new culture mediums that able to maintain or differentiate into the desired cell type with high throughput and also diminish the risk of causing cancer is quite difficult. Afterwards, the laboratories and the companies will need to implement several protocols to work under good manufacturing practice guidelines (GMP) and follow specific storage rules of products that want to be used as a therapy. Several delicate conditions need to be fulfilled for these cells, to be approved by health authorities, and to be used in humans and in clinical trials.

First, we want to review the different cell types, their potential sources, and characteristics and underlie why mainly mesenchymal stem/stromal cells are used for the treatment of sepsis and ARDS. Some pre-clinical studies using induced pluripotent stem cells (iPSC) for the treatment of both syndromes and alveolar type II cells for ARDS treatment also presented some interesting results. The different cell subsets are summarized in **Table 1.**

2.1 Embryonic stem cells (ESCs)

ESCs are pluripotent cells derived from the inner blastocyst cell mass and constitute a potentially unlimited source of cells that could be differentiated into any progenitor cell and used in the clinical trials. ESCs have high plasticity and theoretically unlimited capacity for self-renewal; ESCs have been suggested for regenerative medicine and tissue replacement; however, their embryologic origin is linked to significant ethical issues regarding the use of these cells [17–19].

2.2 Induced pluripotent stem cells (iPSC)

A new type of pluripotent cells, iPSCs, can be obtained by reprogramming animal and human somatic (differentiated) cells. Usually, iPSCs are obtained from dermal fibroblasts because it is an easy source and did not produce any damage to the donor when we obtain them. The cells should be dedifferentiated following

Cell type	Harvest method	Advantages	Disadvantages	Benefits in ARDS	Benefits in sepsis
ESC	Embryos	Totipotent	High tumorigenic potential. Ethical problem	Easy to differentiate to AEC2. No tested *in vivo*	Reduces mortality and decreases lung inflammation
iPSC	Skin biopsy	Easy isolation No rejection	High tumorigenic potential	Easy to differentiate to AEC2. No tested *in vivo*	—
MSC	Bone marrow or adipose tissue	Easy isolation No rejection	High tumorigenic potential	Immunomodulatory effect. Reduces inflammation and lung edema	Reduces mortality and inflammation. Antibacterial activity and antiapoptotic activity
EnPC	Blood	Nontumorigenic	Difficult isolation and small amount	Maintains the integrity of the lung and improves the lung function	Reduces the sepsis damage re-establishing micro and macrocirculation
EpPC	Donor tissue	Nontumorigenic	Difficult isolation and small amount	AEC2 cells were tested improving lung function and reducing inflammation	—

Table 1.
Summary of the cell sources and their benefits.

reprogramming and finally are able to express four transcription factors such as Ocet3/4, Sox2, Klf-4, and c-Myc. They grow indefinitely and differentiate into all cell types of the human body, and also they can be obtained in autologous way reducing the graft-versus-host disease [20–22].

The main problem is that the dedifferentiation and reprogramming have low efficiency, and their genomic modification could create associated problems that until now are unknown or have not been studied and have a high tumorigenicity risk [23–25]. The potential applications of iPSCs in sepsis and ARDS are many [26, 27].

2.3 Endothelial progenitor cells (EnPC)

The endothelial damage is one of the main hallmarks of ARDS, and EnPCs have the ability to regenerate endothelial cells and could have an unlimited role in repairing the damaged endothelium. EnPCs have been defined as circulating cells that have ability to adhere to the endothelium at sites of hypoxia and ischemia secreting pro-angiogenic factors and generate a new vessel [28].

EnPCs express hematopoietic surface cell markers such as CD34. However, the role, isolation, and identification of these cells are not completely elucidated.

EnPCs can be useful as a regenerative instrument to treat several vascular diseases, but the ability to adhere to the endothelium at sites of hypoxia and ischemia secretes pro-angiogenic factors and generates a new vessel [29, 30].

Very few pre-clinical studies have been published, which have used EnPCs in sepsis and ARDS [31–34].

2.4 Epithelial progenitor cells (EpPC)

Some epithelial progenitors of the lung alveolar compartment have also been identified, but their isolation is really difficult and the number of cells that could be obtained is really low. EpPCs might be useful in the treatment of ARDS.

EpPCs are specified during development in each tissue and are highly regulated by epithelial-mesenchymal interactions. Different tissues have different EpPCs with small adjustments in their function; however, their maintenance, activation, and differentiation are regulated by the same pathways between all the tissues [35, 36].

The use of alveolar-epithelial type II cells (AE2C) for the treatment of ARDS has been demonstrated; these cells are more differentiated than EpPCs, but still can proliferate and differentiate into alveolar-epithelial type I cell, that are the complete differentiated cells in the lung epithelia. AE2C cells can only be isolated from the lungs of organ donors and may have problems in graft-versus-host disease [37–39]. The isolation of AE2C is laborious, but they have less ethical problems and tumorigenicity potential.

2.5 Mesenchymal stem cells (MSC)

MSCs are the best described cells and mostly used as a cell therapy. MSCs are multipotent cells that have been isolated from several tissues such as umbilical cord blood, placenta, adipose tissue, lung, and bone marrow.

The International Society of Cellular Therapy defined that MSCs should follow the three criteria: (1) MSCs must be adherent to plastic; (2) MSCs must express some cell surface markers, such as CD105, CD90, and CD73, but must not express other markers, including CD45, CD34, CD14, or CD11b; and (3) MSCs must have the capacity to differentiate into mesenchymal lineages (osteoblasts, adipocytes, and chondroblasts) in *in vitro* conditions [40].

MSCs have a high degree of plasticity and can be differentiated into a variety of cell lineages, but they do not possess the complete plasticity of ESCs. However, MSCs have some advantages because of their easy isolation and enormous propagation in culture and also because their use does not involve the ethical problems associated to the use of ESCs [41]. Moreover, they can be obtained in autologous way diminishing the immune rejection problem. In addition, MSCs are not immunogenic; they have an innate ability to avoid detection by a recipient's immune system because they express intermediate levels of major histocompatibility class I but do not express major histocompatibility class II [42–44].

Several experimental studies have indicated that MSCs may have potential therapeutic application in sepsis and ARDS [45–49]. It has also been reported that MSCs release several microvesicles that might have the therapeutic potential [50, 51].

3. Mesenchymal stem cell therapy for sepsis and acute respiratory distress syndrome

The complex pathophysiology of sepsis and also of ARDS requires a therapy with a wide range of properties. The heterogeneity of the host response in front of sepsis or ARDS makes it difficult to find a proper drug that works as a therapy. Stem cells are potential therapeutic agents with a diverse spectrum of action and can act at different levels of the pathophysiology of ARDS or sepsis, non only treating inflammation, moreover affecting coagulation, enhancing antimicrobial effect, modulating the innate and adaptive immune response, and reducing endothelium permeability and edema.

MSCs can be administered intravenously, but then only a small percentage of injected cells arrive to the injury site. Usually, MSCs get trapped into the lung's microvasculature, and after some hours, they are engulfed by macrophages and they disappear. It is well known that cells do not engraft into the tissue and their primary action mechanism is the secretion of soluble factors with therapeutic properties.

MSCs release into the extracellular media bioactive cytokines, chemokines, angiogenic factors, and/or growth factors. Moreover, it has been described that MSCs also release extracellular vesicles with bioactive compounds. Extracellular vesicles are small vesicles made by a phospholipid bilayer and encapsulate proteins, lipids, or miRNAs or other compounds that are protected from the media and can be phagocytosed by other cells and act on them. Nowadays, the use of extracellular vesicles to target specific cells is also a raising up therapy competing with cell therapies; however, there is still a lot to define about the secretome of the MSC to be able to mimic its therapeutic effects.

All these factors directly secreted to the media or inside extracellular vesicles regulate intracellular pathways from different cells and can act on the innate and adaptive immune system. The different effects or mechanisms are described in the following sections and in **Figure 1**.

3.1 Effects on the innate immune system

Inflammation is one of the main drivers of ARDS and sepsis pathogenesis. During ARDS, the injury of the lung endothelium and epithelium is producing a recruitment of pro-inflammatory cells such as neutrophils and monocytes into the alveolar space. These pro-inflammatory cells release several pro-inflammatory cytokines, for example, TNF-α, IL-1β, and IL-6 [52–54]. Moreover, neutrophils are secreting several ROS enhancing the damage of the endothelium and epithelial layers propagating the damage [55–57].

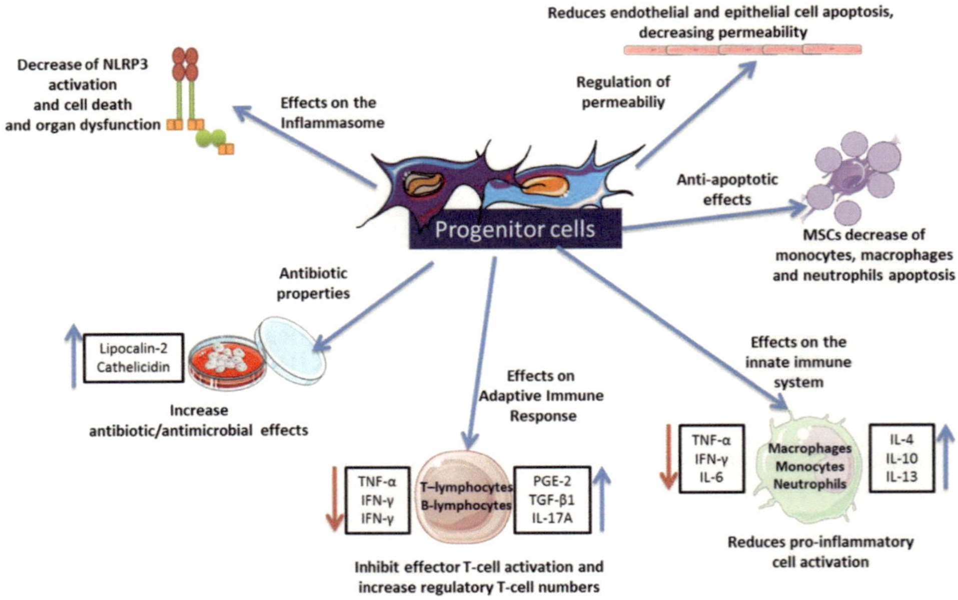

Figure 1.
Potential mechanisms of the cell therapies.

During sepsis, the infection activates TLR4, which leads to the activation of MyD88 and NF-kb and the transcription of pro-inflammatory cytokines such as TNF-α, IL-1β, and IL-6. After the first hours following the infection, there is a specific storm of cytokines [58].

MSCs have the ability to modulate the immune response and secrete several anti-inflammatory cytokines such as IL-4, IL-10, or IL-13 [59]. Furthermore, they regulate inflammation using different strategies and acting on different cells, which are described in the following sections.

3.1.1 Effects on humoral immune response

Numerous studies proved that MSCs decrease the pro-inflammatory cytokine response (TNF-α, IFN-γ, and IL-1α, -1β, -6, -12, and -17). They promote this effect while increasing concentrations of the anti-inflammatory agents, including IL-1 receptor antagonist (IL-1Ra), IL-10, tumor necrosis factor-inducible gene 6 protein (TSG-6), cyclooxygenase-2 (COX-2), insulin-like growth factor 1 (IGF-1), and prostaglandin-E2 (PGE2) [60, 61].

All these factors act on different cells; for example, IL-1Ra is able to reduce the production of inflammatory TNF-α in macrophages and also inhibits helper-T lymphocytes activation. TSG-6 has a potent anti-inflammatory effect acting in macrophages and polarizing them from a pro- to an anti-inflammatory phenotype while reducing the secretion of inflammatory chemokines by these cells. IGF-1 has been described as an anti-apoptotic compound, and during sepsis and ARDS, less apoptosis of the endothelium and epithelium cells reduces the damage and the associated inflammation [62, 63].

3.1.2 Effects on the inflammasome

The inflammasome is a multiprotein intracellular oligomer that detects pathogenic microorganisms and other factors and activates an inflammatory response. Inflammasome activates caspase-1 and caspase-1 cleaves the precursor cytokines pro-IL-1β and pro-IL-18, generating the biologically active cytokines IL-1β and IL-18, respectively. Caspase 1 also activates following compounds derived from the host innate immune system that can lead to pyropotosis (an inflammatory form of cell death) [64].

Generally, myeloid cells such as neutrophils, monocytes, and macrophages are expressing inflammasome proteins such as pattern-recognition receptors (PRRs), toll-like receptors (TLR), or C-type lectin receptors in their membrane. NLRP3 is the best characterized and versatile inflammasome and is mainly expressed by myeloid lineage cells and stimulated by the activation of TLR and other signals [64, 65].

MSCs can regulate the NLRP3 inflammasome activation through the secretion of PGE2 that leads to an increase in IL-10 production by macrophages. The decrease in NLRP3 activation moderates cell death and organ dysfunction [65].

3.1.3 Effects on neutrophil response

Sepsis is associated with neutropenia, and it has been shown that MSCs are able to increase the neutrophil counts after their administration [66, 67]. The activity and survival of neutrophils were also increased in the lung during ARDS; this effect is produced through the stimulation of TLR-3 by MSCs.

During sepsis and ARDS, the injury or pathogen activates some recognition receptors that promote activated neutrophils recruitment. Neutrophils release some antimicrobial compounds and produce some traps to eliminate the pathogen. During sepsis, the general infection produces an overactivation of the neutrophils that migrate to inflamed and noninflamed tissue and can lead to an organ dysfunction [68].

Several pre-clinical studies have demonstrated that MSCs are able to modify neutrophils behavior, maintaining their bactericidal function but reducing the host injury.

In pre-clinical sepsis models, MSC therapy diminishes neutrophil infiltration into several organs such as lung, liver, gut, and kidney, reducing injury and improving organ function. Besides, MSCs augment neutrophil-mediated phagocytosis, improving the clearance of bacteria. It has been demonstrated that the protective effect of MSCs in systemic sepsis is clearly mediated by neutrophils, because their depletion abolishes the protective effect of MSCs [55, 69, 70].

3.1.4 Effects on monocyte/macrophage response

Monocytes and macrophages are present in practically all tissues. They have a wide range of functions like maintenance of the tissue homeostasis, immunologic functions, and participation in several metabolic pathways [71, 72].

It is well known that, in the course of sepsis and ARDS, macrophages get dysfunction and are not able to perform their activity as usual. Several groups focus in the use of MSCs as modulator of macrophages activity [73, 74]. Tissue resident macrophages and also circulating monocytes that migrate into the tissue and convert to macrophages can polarize to different phenotypes. Pro-inflammatory or M1 macrophages produce several pro-inflammatory cytokines and are involved in the elimination of pathogens. Anti-inflammatory/reparative or M2 macrophages secrete more anti-inflammatory compounds and are involved in the clearance of apoptotic cells. Alteration of M1 macrophages to a M2 phenotype has been demonstrated to be important to damage resolution.

MSCs have the ability to secrete several factors that polarize macrophages to a M2 phenotype, promoting a resolution phase, increasing phagocytic activity, and decreasing inflammation. It has been described that macrophages secrete PGE-2 that is able to increase the production of IL-10, SOCS-3, TGF-β, TSG-6, and others [60, 75]. All these factors are able to reduce the recruitment/migration of pro-inflammatory cells into the tissue, preventing organ dysfunction and also decreasing the production of pro-inflammatory cytokines such as TNF-α or IFN-γ by macrophages [76].

Furthermore, complement activation pathway is also upregulated by MSCs infusion, what leads to a more efficient clearance of pathogens. MSCs, as we had previously described, secrete KGF that also promotes a M2 phenotype and can transfer mitochondria to macrophages, also reducing their pro-inflammatory phenotype [73, 77]. During ARDS, it has been shown that MSCs attenuate the damage induced by bacteria or LPS through the inhibition of Wnt/β-catenin pathway. The effects of MSCs on the macrophage activation varies quite a lot depending on the organ, the stage of the disease, and the steady-state of these macrophages when MSCs are infused.

3.2 Effects on adaptive immune response

In several diseases, there is an enhanced activation and proliferation of T and B cells, and this is also happening during sepsis. MSCs have the ability to diminish this cell activity and their proliferation [78, 79].

Explicitly, it has been shown that MSCs inhibit effector T-cell activation and can increase regulatory T-cell numbers, while suppressing propagation of CD4+ T-helper cells, CD8+ cytotoxic T lymphocytes, and natural killer cells [80, 81]. This effect is mediated through the secretion of PGE-2 and TGF-β1 [82]. Moreover, MSCs promote the formation of CD8+ regulatory T cells that might decrease cytotoxicity caused by cytotoxic lymphocytes [83, 84].

Regulatory T cells are a subclass of T cells, which functions in modulating the immune system and maintaining the antigen tolerance. These cells are able to limit inflammation and reduce organ dysfunction in sepsis, and MSCs have the capacity to regulate regulatory-T cell function [81, 85]. It has been verified that regulatory T cells subset is necessary to eliminate bacteria during infections.

MSCs induce regulatory T cells promoting their efficiency and enhancing sepsis or ARDS resolution [86, 87].

It has also been reported that MSCs modified the activity of other cells from the adaptive immune system such as NK, regulatory B cells, and dendritic cells.

3.3 Antibiotic properties

MSCs have been reported to have antibiotic/antimicrobial effects. They reduced bacterial levels in bronchoalveolar lavage, blood, spleen, and lung tissue. Generally, the antimicrobial effect of MSCs is due to their effect on host immune cells. MSCs have the ability to increase the phagocytic capacity of the host immune cells such as macrophages, monocytes, dendritic cells, and neutrophils [54, 73, 88]. It has been shown that this effect is produced through the production of keratinocyte growth factor (KGF) or also named fibroblast growth factor 7 (FGF7) [62, 69, 89].

Besides, MSCs are producing other factors with antibiotic properties per se. For example, in some pre-clinical models, it has been described that mouse MSCs secrete lipocalin-2, also known as neutrophil gelatinase-associated lipocalin (NGAL), which limits bacterial growth by iron sequestration. Human MSCs secrete LL-37, also known as cathelicidin antimicrobial peptide 18, described as antimicrobial peptide [90]. Patients with a high level of LL-37 were more likely to survive to a strong infection.

3.4 Antiapoptotic effects

During sepsis and ARDS, the apoptosis of endothelial, epithelial, and immune cells is one of the main descriptors of the severity of the disease. MSC therapy has been confirmed that it is able to limit the apoptosis of host cells.

The antiapoptotic capacity has been tested *in vitro* incubating neutrophils with MSCs or their supernatants and also *in vivo* where resident macrophages have presented less apoptosis [91, 92]. It seems that antiapoptotic effect of MSCs does not require cell-cell contact, and IL-6 and FGF7 have been described as the main drivers of this effect.

The role of MSCs in the decrease of monocytes, macrophages, and neutrophils apoptosis is directly associated with an increase in the clearance of bacteria and also can explain the antibiotic properties of MSCs [69].

3.5 Regulation of permeability

The endothelial and epithelial injury is a crucial characteristic of sepsis and ARDS. Throughout sepsis and ARDS, the barrier function of the endothelium and epithelium is destroyed due to the loss of their integrity and the disruption of the junction proteins between cells [1, 93].

MSCs have been described that they are able to decrease permeability and decrease the disturbance of the membrane, promoting the production of tight junction proteins and limiting the binding of inflammatory cells to the endothelium. It seems from some *in vitro* studies that the preserving effect of MSCs on permeability is due to the secretion of IL-1Ra and PGE-2, which decrease inflammation and reduce endothelial and epithelial cell apoptosis [94–96].

4. Mechanisms by which MSCs exert their effects

MSCs work by multiple mechanisms and can exert their effect through cell-cell contact, secreting several factors directly to the media or through the release of extracellular vesicles.

It seems that during sepsis and ARDS the MSC are not really engrafted in any tissue; however, it has been demonstrated that cells migrate to the site of the injury and they are retained there for a while. It has been shown that some effects are produced through cell-cell contact between MSCs and alveolar epithelial cells mainly to regulate endothelial integrity creating some junctions and transferring mitochondria or other cellular products with therapeutic effect [58, 60, 89].

In the last sections, we reviewed the effects of MSCs through the secretion of several factors such as antimicrobial peptides, antiapoptotic effectors, or immunomodulatory mediators.

Besides, MSCs release extracellular vesicles, which encapsulate several cellular components, including mitochondria and gene products such as miRNAs and mRNAs. Moreover, it has been described that some extracellular vesicles can also encapsulate lipids and proteins. Several studies have described the delivery of miR-223 that is transfer to macrophages and cardiomyocytes and reduces their inflammatory response. Also, the mRNA from KGF has also been detected inside these extracellular vesicles, producing its effect in the endothelium and epithelium and enhancing their repair [97–99].

5. Clinical trials

The encouraging preclinical data suggest that cell-based therapies capable of simultaneously affecting multiple processes constitute a promising new approach to sepsis and ARDS treatment [100]. MSCs can be efficiently cultured from bone marrow, umbilical cord blood, adipose tissue, and other sources and have a low

Study title	Register number	Sepsis/ARDS	Cell type	Phase	Dose and via	Start date	Finish date	Country	Reference
Adipose-derived mesenchymal stem cells in acute respiratory distress syndrome	NCT01902082	ARDS	MSC	Phase 1	1×10^6 intravenous	November 2012	November 2014	China	[104]
Russian clinical trial of mesenchymal cells in patients with septic shock and severe neutropenia	NCT01849237	Sepsis	MSC	Phase 1/2	1–2 millions/kg/day intravenous	December 2012	May 2015	Russia	[109]
Human mesenchymal stem cells for acute respiratory distress syndrome (START)	NCT01775774 NCT02097641	ARDS	MSC	Phase 1/2a	1 or 5 or 10 million cells/kg	July 2013	February 2018	USA	[102, 108]
Cellular immunotherapy for septic shock: a phase I trial (CISS)	NCT02421484	Sepsis	MSC	Phase 1	0.3 or 1 or 3 million cells/kg	May 2015	June 2017	Canada	[103]
Treatment of severe acute respiratory distress syndrome with allogeneic bone marrow-derived mesenchymal stromal cells	NCT02215811	ARDS	MSC	Phase 1	Not known. Cells are combined with ECMO	March 2014	December 2015	Sweden	
Human umbilical cord-derived mesenchymal stem cell therapy in acute lung injury (UCMSC-ALI)	NCT02444455	ARDS	MSC	Phase 1/2	5×10^5/kg intravenous	May 2015	December 2017	China	
A phase 1/2 study to assess MultiStem® therapy in acute respiratory distress syndrome (MUST-ARDS)	NCT02611609	ARDS	Multistem (MSC)	Phase 1/2	Not known	January 2016	November 2018	USA/UK	
Effects of administration of mesenchymal stem cells on organ failure during the septic shock (CSM choc)	NCT02883803	Sepsis	MSC	Phase 1	10^6/kg intravenous	December 2016	December 2019	France	
Mesenchymal stem cells (MSCs) for treatment of acute respiratory distress syndrome (ARD) in stem cell transplant patients	NCT02804945	ARDS	MSC	Phase 2	3×10^6 cell/kg intravenous	February 2017	February 2019	USA	
Repair of acute respiratory distress syndrome by stromal cell administration (REALIST)	NCT03042143	ARDS	MSC	Phase 1/2	2 doses, not specified	September 2017	September 2020	UK	

Table 2.
Phase 1 and phase 2 clinical studies with cell therapies for ARDS and sepsis (adapted from reference X).

expression of major histocompatibility antigens, permitting allogeneic therapy without need for immunosuppression, and no safety issues have been identified in hundreds of patients [101].

Two recent phase 1 dose-escalation safety of MSCs in patients with ARDS [102] and septic shock [103] raised no safety concerns adding a growing body of evidence that MSCs can be safety administered intravenously at several millions cells per kilogram to critically ill patients. Zheng et al. assigned 12 patients with moderate-to-severe ARDS to receive 1 million adipose-derived MSCs/kg or saline, reporting no adverse infusion-related events [104]. Nevertheless, the intravenous administration of high doses of MSCs may be associated with vascular thrombosis. Recently published trials of MSCs in ARDS and in sepsis and studies in progress are summarized in **Table 2**.

Larger phase 2 and phase 3 trials are required to define the adverse events related to cell therapy. These future studies should be designed considering the use of fresh versus cryopreserved cell product, the optimal dosing and route administration regimen, and which biologic and clinical outcomes should be assessed for insights into safety and efficacy. Central challenges in ensuring consistency in final product in the cell therapy field are the lack of consensus for *in vitro* potency assays to optimize donor selection, MSC tissue source, and optimal MSC culture condition. Without an accepted potency assay, the lack of a signal of biological effect in patients is difficult to interpret. One approach may be to measure paracrine factors such as angiopoietin-1 and KGF. Phase 2 trials should identify which biological markers are altered by cell-based therapy. However, the optimal approach for patient selection in sepsis and ARDS trials remains a challenge. ARDS and sepsis are clinical syndromes rather than a disease, with a lack of specificity of clinical criteria. Thirty-five percent of patients with ARDS have a hyperinflammatory phenotype, associated with a higher mortality and a different therapeutic response. Patients with a hyperinflammatory phenotype might be better candidates for therapy with MSCs [105, 106].

There is a growing evidence for the therapeutic effects of extracellular vesicles from MSCs, raising the possibility that cell-free therapy consisting of exosomes or microvesicles or MSC culture media might be produced and could be tested in patients with sepsis, ARDS, acute kidney injury, or traumatic brain injury. Several steps would be needed for this approach to become a reality, including optimization of purification methods for isolation of the required fractions of extracellular vesicles from MSCs accompanied by a comprehensive characterization of RNA, microRNA, lipids, and proteins in exosomes or microvesicles [50].

Although cell donors are extensively screened to rule out systemic illnesses, other donor-related variables, such as age, may be important. MSCs from aging (murine) donors demonstrated reduced efficacy [107]. Variations in production and cryopreservation methods may impact variability in the function of MSCs when tested in preclinical models or in patients for specific clinical disorders. Dimethyl sulfoxide and cell debris are removed by centrifugation after the cryopreserved MSCs have been thawed and cells are suspended in plasmalyte before intravenous administration in preclinical studies of lung injury. Current clinical studies are focused on a single dose of MSC administration via the less-invasive intravenous route with the START trial demonstrating safety for this approach [108].

6. Conclusions

Extensive progress has been made in the last years concerning cell therapy for sepsis and ARDS. Cell therapies have shown promising results in pre-clinical studies. Several pathways, proteins, miRNAs, and lipids have been characterized and explain the mechanism of action of these cell therapies.

Several questions need further study, including determining the best source for the MSCs isolation, their large-scale production, and cryopreservation. Moreover, the therapeutic potential of MSCs and its conditioned media need to be studied for checking their efficacy in short-term and long-term follow-up studies.

The heterogeneity of patients with sepsis and ARDS is enormous, and establish a target population or the stratification of the patients will help us to determine better the therapeutic effect of these therapies.

There are many complications and concerns with using stem cells for cell-based therapy. The future may emphasis on the stimulation of other cells (growth factors, cytokines, and various other hematopoietic elements) that facilitate the formation or repair of endothelium and epithelium and the modulation of inflammatory cells.

We need to await evidence that these cell therapies have a benefit in patients with sepsis or ARDS and evaluate the phase I and II results from the ongoing studies.

Acknowledgements

This work was supported by Ministerio de Economía y Competitividad-Instituto de Salud Carlos III (PI12/02548) Cofinanciado por el Fondo Europeo de Desarrollo Regional (FEDER), SEPAR (2016/096) and CIBER de Enfermedades Respiratorias (CIBERES).

Conflict of interest

The authors declare that they have no conflict of interest.

Abbreviation list

AE2C	alveolar-epithelial type II cells
ARDS	acute respiratory distress syndrome
COX-2	cyclooxygenase-2
EnPC	endothelial progenitor cells
EpPC	epithelial progenitor cells
ESC	embryonic stem cells
FGF7	fibroblast growth factor 7
GMP	good manufacturing practice guidelines
IGF-1	insulin-like growth factor 1
IL-1Ra	IL-1 receptor antagonist
iPSC	induced pluripotent stem cells
MSC	mesenchymal stem/stromal cells
NGAL	neutrophil gelatinase-associated lipocalin
PEEP	positive end-expiratory pressure
PGE2	prostaglandin-E2
PRRs	pattern-recognition receptors
TLR	toll like receptors
TSG-6	tumor necrosis factor-inducible gene 6 protein

Author details

Raquel Guillamat-Prats[1,2]* and Antonio Artigas[1,2,3,4]

1 Institut d'Investigació i Innovació Parc Tauli (I3PT), Sabadell, Spain

2 Centro de Investigaciones Biomédicas en Red de Enfermedades Respiratorias (CIBERES), Madrid, Spain

3 Universitat Autònoma de Barcelona, Bellaterra, Spain

4 Critical Care Center, Corporació Sanitària Universitaria Parc Taulí, Sabadell, Spain

*Address all correspondence to: r.guillamat.prats@gmail.com

References

[1] Angus DC, van der Poll T. Severe sepsis and septic shock. The New England Journal of Medicine. 2013;**369**:2063

[2] Levy MM, Fink MP, Marshall JC, Abraham E, Angus D, Cook D, et al. 2001 SCCM/ESICM/ACCP/ATS/SIS international sepsis definitions conference. Intensive Care Medicine. 2003;**29**:530-538

[3] Matthay MA, Ware LB, Zimmerman GA. The acute respiratory distress syndrome. The Journal of Clinical Investigation. 2012;**122**:2731-2740

[4] Ware LB, Matthay MA. The acute respiratory distress syndrome. The New England Journal of Medicine. 2000;**342**:1334-1349

[5] Freund Y, Lemachatti N, Krastinova E, Van Laer M, Claessens Y-E, Avondo A, et al. Prognostic accuracy of Sepsis-3 criteria for in-hospital mortality among patients with suspected infection presenting to the emergency department. Journal of the American Medical Association. 2017;**317**:301-308

[6] Martin-Loeches I, Levy MM, Artigas A. Management of severe sepsis: Advances, challenges, and current status. Drug Design, Development and Therapy. 2015;**9**:2079-2088

[7] Luo L, Shaver CM, Zhao Z, Koyama T, Calfee CS, Bastarache JA, et al. Clinical predictors of hospital mortality differ between direct and indirect ARDS. Chest. 2017;**151**:755-763

[8] Rubenfeld GD, Caldwell E, Peabody E, Weaver J, Martin DP, Neff M, et al. Incidence and outcomes of acute lung injury. The New England Journal of Medicine. 2005;**353**:1685-1693

[9] Martínez ML, Ferrer R, Torrents E, Guillamat-Prats R, Gomà G, Suárez D, et al. Impact of source control in patients with severe sepsis and septic shock. Critical Care Medicine. 2017;**45**:11-19

[10] Ferrer R, Martínez ML, Gomà G, Suárez D, Álvarez-Rocha L, de la Torre MV, et al. Improved empirical antibiotic treatment of sepsis after an educational intervention: The ABISS-Edusepsis study. Critical Care. 2018;**22**:167

[11] Narendra DK, Hess DR, Sessler CN, Belete HM, Guntupalli KK, Khusid F, et al. Update in management of severe hypoxemic respiratory failure. Chest. 2017;**152**:867-879

[12] Patroniti N, Bonatti G, Senussi T, Robba C. Mechanical ventilation and respiratory monitoring during extracorporeal membrane oxygenation for respiratory support. Annals of Translational Medicine. 2018;**6**:386

[13] Alessandri F, Pugliese F, Ranieri VM. The role of rescue therapies in the treatment of severe ARDS. Respiratory Care. 2018;**63**:92-101

[14] Kallet RH. A comprehensive review of prone position in ARDS. Respiratory Care. 2015;**60**:1660-1687

[15] Horie S, Gonzalez HE, Laffey JG, Masterson CH. Cell therapy in acute respiratory distress syndrome. Journal of Thoracic Disease. 2018;**10**:5607-5620

[16] Gotts JE, Matthay MA. Cell-based therapy in sepsis. A step closer. American Journal of Respiratory and Critical Care Medicine. 2018;**197**:280-281

[17] Martin GR, Evans MJ. Differentiation of clonal lines of teratocarcinoma cells: Formation of embryoid bodies in vitro. Proceedings of the National Academy of Sciences of the United States of America. 1975;**72**:1441-1445

[18] Thomson JA, Itskovitz-Eldor J, Shapiro SS, Waknitz MA, Swiergiel JJ, Marshall VS, et al. Embryonic stem cell lines derived from human blastocysts. Science. 1998;**282**:1145-1147

[19] Toya SP, Li F, Bonini MG, Gomez I, Mao M, Bachmaier KW, et al. Interaction of a specific population of human embryonic stem cell-derived progenitor cells with CD11b+ cells ameliorates sepsis-induced lung inflammatory injury. The American Journal of Pathology. 2011;**178**:313-324

[20] Yu J, Thomson JA. Pluripotent stem cell lines. Genes & Development. 2008;**22**:1987-1997

[21] Takahashi K, Yamanaka S. Induction of pluripotent stem cells from mouse embryonic and adult fibroblast cultures by defined factors. Cell. 2006;**126**:663-676

[22] Okita K, Ichisaka T, Yamanaka S. Generation of germline-competent induced pluripotent stem cells. Nature. 2007;**448**:313-317

[23] Ghaedi M, Mendez JJ, Bove PF, Sivarapatna A, Raredon MSB, Niklason LE. Alveolar epithelial differentiation of human induced pluripotent stem cells in a rotating bioreactor. Biomaterials. 2014;**35**:699-710

[24] Ghaedi M, Calle EA, Mendez JJ, Gard AL, Balestrini J, Booth A, et al. Human iPS cell-derived alveolar epithelium repopulates lung extracellular matrix. The Journal of Clinical Investigation. 2013;**123**:4950-4962

[25] Wang C, Hei F, Ju Z, Yu J, Yang S, Chen M. Differentiation of urine-derived human induced pluripotent stem cells to alveolar type II epithelial cells. Cellular Reprogramming. 2016;**18**:30-36

[26] Scudellari M. How iPS cells changed the world. Nature. 2016;**534**:310-312

[27] Heffernan C, Sumer H, Verma PJ. Generation of clinically relevant "induced pluripotent stem" (iPS) cells. Journal of Stem Cells. 2011;**6**:109-127

[28] Timmermans F, Plum J, Yöder MC, Ingram DA, Vandekerckhove B, Case J. Endothelial progenitor cells: Identity defined? Journal of Cellular and Molecular Medicine. 2009;**13**:87-102

[29] Hristov M, Erl W, Weber PC. Endothelial progenitor cells: Mobilization, differentiation, and homing. Arteriosclerosis, Thrombosis, and Vascular Biology. 2003;**23**:1185-1189

[30] Yoder MC. Human endothelial progenitor cells. Cold Spring Harbor Perspectives in Medicine. 2012;**2**:a006692

[31] Cao J-P, He X-Y, Xu H-T, Zou Z, Shi X-Y. Autologous transplantation of peripheral blood-derived circulating endothelial progenitor cells attenuates endotoxin-induced acute lung injury in rabbits by direct endothelial repair and indirect immunomodulation. Anesthesiology. 2012;**116**:1278-1287

[32] Li H, Qiang Y, Wang L, Wang G, Yi J, Jing H, et al. Repair of lipopolysaccharide-induced acute lung injury in mice by endothelial progenitor cells, alone and in combination with simvastatin. Chest. 2013;**144**:876-886

[33] Lam C-F, Liu Y-C, Hsu J-K, Yeh P-A, Su T-Y, Huang C-C, et al. Autologous transplantation of endothelial progenitor cells attenuates acute lung injury in rabbits. Anesthesiology. 2008;**108**:392-401

[34] Xu X, Yang J, Li N, Wu R, Tian H, Song H, et al. Role of endothelial progenitor cell transplantation in rats with sepsis. Transplantation Proceedings. 2015;**47**:2991-3001

[35] Blanpain C, Horsley V, Fuchs E. Epithelial stem cells: Turning over new leaves. Cell. 2007;**128**:445-458

[36] Rawlins EL. Lung epithelial progenitor cells: Lessons from development. Proceedings of the American Thoracic Society. 2008;**5**:675-681

[37] Guillamat-Prats R, Puig F, Camprubí-Rimblas M, Herrero R, Serrano-Mollar A, Gómez MN, et al. Intratracheal instillation of alveolar type II cells enhances recovery from acute lung injury in rats. The Journal of Heart and Lung Transplantation. 2018;**37**:782-791

[38] Adamson IY, Bowden DH. Derivation of type 1 epithelium from type 2 cells in the developing rat lung. Laboratory Investigation: A Journal of Technical Methods And Pathology. 1975;**32**:736-745

[39] Evans MJ, Cabral LJ, Stephens RJ, Freeman G. Transformation of alveolar type 2 cells to type 1 cells following exposure to NO_2. Experimental and Molecular Pathology. 1975;**22**:142-150

[40] Dominici M, Le Blanc K, Mueller I, Slaper-Cortenbach I, Marini F, Krause D, et al. Minimal criteria for defining multipotent mesenchymal stromal cells. The International Society for Cellular Therapy position statement. Cytotherapy. 2006;**8**:315-317

[41] Pourrajab F, Forouzannia SK, Tabatabaee SA. Molecular characteristics of bone marrow mesenchymal stem cells, source of regenerative medicine. International Journal of Cardiology. 2013;**163**:125-131

[42] Bianco P, Cao X, Frenette PS, Mao JJ, Robey PG, Simmons PJ, et al. The meaning, the sense and the significance: Translating the science of mesenchymal stem cells into medicine. Nature Medicine. 2013;**19**:35-42

[43] Salmikangas P, Menezes-Ferreira M, Reischl I, Tsiftsoglou A, Kyselovic J, Borg JJ, et al. Manufacturing, characterization and control of cell-based medicinal products: Challenging paradigms toward commercial use. Regenerative Medicine. 2015;**10**:65-78

[44] Deskins DL, Bastakoty D, Saraswati S, Shinar A, Holt GE, Young PP. Human mesenchymal stromal cells: Identifying assays to predict potency for therapeutic selection. Stem Cells Translational Medicine. 2013;**2**:151-158

[45] Curley GF, Hayes M, Ansari B, Shaw G, Ryan A, Barry F, et al. Mesenchymal stem cells enhance recovery and repair following ventilator-induced lung injury in the rat. Thorax. 2012;**67**:496-501

[46] Chimenti L, Luque T, Bonsignore MR, Ramírez J, Navajas D, Farré R. Pre-treatment with mesenchymal stem cells reduces ventilator-induced lung injury. The European Respiratory Journal. 2012;**40**:939-948

[47] Curley GF, Ansari B, Hayes M, Devaney J, Masterson C, Ryan A, et al. Effects of intratracheal mesenchymal stromal cell therapy during recovery and resolution after ventilator-induced lung injury. Anesthesiology. 2013;**118**:924-932

[48] Sepúlveda JC, Tomé M, Fernández ME, Delgado M, Campisi J, Bernad A, et al. Cell senescence abrogates the therapeutic potential of human mesenchymal stem cells in the lethal endotoxemia model. Stem Cells. 2014;**32**:1865-1877

[49] Gonzalez-Rey E, Anderson P, González MA, Rico L, Büscher D, Delgado M. Human adult stem cells derived from adipose tissue protect against experimental colitis and sepsis. Gut. 2009;**58**:929-939

[50] Matthay MA. Extracellular vesicle transfer from mesenchymal stromal cells modulates macrophage function in acute lung injury: Basic science and clinical implications.

American Journal of Respiratory and Critical Care Medicine. 15 Nov 2017;**196**(10):1234-1236

[51] Zheng G, Huang R, Qiu G, Ge M, Wang J, Shu Q, et al. Mesenchymal stromal cell-derived extracellular vesicles: Regenerative and immunomodulatory effects and potential applications in sepsis. Cell and Tissue Research. 2018;**374**:1-15

[52] Mao M, Wang S-N, Lv X-J, Wang Y, Xu J-C. Intravenous delivery of bone marrow-derived endothelial progenitor cells improves survival and attenuates lipopolysaccharide-induced lung injury in rats. Shock. 2010;**34**:196-204

[53] Gupta N, Su X, Popov B, Lee JW, Serikov V, Matthay MA. Intrapulmonary delivery of bone marrow-derived mesenchymal stem cells improves survival and attenuates endotoxin-induced acute lung injury in mice. The Journal of Immunology. 2007;**179**:1855-1863

[54] Luo C, Zhang F, Zhang L, Geng Y, Li Q, Hong Q, et al. Mesenchymal stem cells ameliorate sepsis-associated acute kidney injury in mice. Shock. 2014;**41**:123-129

[55] Hall SRR, Tsoyi K, Ith B, Padera RF, Lederer JA, Wang Z, et al. Mesenchymal stromal cells improve survival during sepsis in the absence of heme oxygenase-1: The importance of neutrophils. Stem Cells. 2013;**31**:397-407

[56] Wang Y, Tan L, Jin J, Sun H, Chen Z, Tan X, et al. Non-cultured dermal-derived mesenchymal cells attenuate sepsis induced by cecal ligation and puncture in mice. Scientific Reports. 2015;**5**:16973

[57] Raffaghello L, Bianchi G, Bertolotto M, Montecucco F, Busca A, Dallegri F, et al. Human mesenchymal stem cells inhibit neutrophil apoptosis: A model for neutrophil preservation in the bone marrow niche. Stem Cells. 2008;**26**:151-162

[58] Xu J, Woods CR, Mora AL, Joodi R, Brigham KL, Iyer S, et al. Prevention of endotoxin-induced systemic response by bone marrow-derived mesenchymal stem cells in mice. American Journal of Physiology. Lung Cellular and Molecular Physiology. 2007;**293**:L131-L141

[59] Rocheteau P, Chatre L, Briand D, Mebarki M, Jouvion G, Bardon J, et al. Sepsis induces long-term metabolic and mitochondrial muscle stem cell dysfunction amenable by mesenchymal stem cell therapy. Nature Communications. 2015;**6**:10145

[60] Németh K, Leelahavanichkul A, Yuen PST, Mayer B, Parmelee A, Doi K, et al. Bone marrow stromal cells attenuate sepsis via prostaglandin E(2)-dependent reprogramming of host macrophages to increase their interleukin-10 production. Nature Medicine. 2009;**15**:42-49

[61] Choi H, Lee RH, Bazhanov N, Oh JY, Prockop DJ. Anti-inflammatory protein TSG-6 secreted by activated MSCs attenuates zymosan-induced mouse peritonitis by decreasing TLR2/NF-κB signaling in resident macrophages. Blood. 2011;**118**:330-338

[62] Liang Z, Sun J, Wang P, Tian Q, Yang Z, Chen L. Bone marrow-derived mesenchymal stem cells protect rats from endotoxin-induced acute lung injury. Chinese Medical Journal. 2011;**124**:2715-2722

[63] Cóndor JM, Rodrigues CE, de Sousa Moreira R, Canale D, Volpini RA, Shimizu MHM, et al. Treatment with human wharton's jelly-derived mesenchymal stem cells attenuates sepsis-induced kidney injury, liver injury, and endothelial dysfunction. Stem Cells Translational Medicine. 2016;**5**:1048-1057

[64] Martinon F, Burns K, Tschopp J. The inflammasome: A molecular platform triggering activation of inflammatory caspases and processing of proIL-beta. Molecular Cell. 2002;**10**:417-426

[65] Miao C-M, Jiang X-W, He K, Li P-Z, Liu Z-J, Cao D, et al. Bone marrow stromal cells attenuate LPS-induced mouse acute liver injury via the prostaglandin E 2-dependent repression of the NLRP3 inflammasome in Kupffer cells. Immunology Letters. 2016;**179**:102-113

[66] Kolaczkowska E, Kubes P. Neutrophil recruitment and function in health and inflammation. Nature Reviews Immunology. 2013;**13**:159-175

[67] Sônego F, Castanheira FVES, Ferreira RG, Kanashiro A, Leite CAVG, Nascimento DC, et al. Paradoxical roles of the neutrophil in sepsis: Protective and deleterious. Frontiers in Immunology. 2016;7:155

[68] Nourshargh S, Renshaw SA, Imhof BA. Reverse migration of neutrophils: Where, when, how, and why? Trends in Immunology. 2016;**37**:273-286

[69] Mei SHJ, Haitsma JJ, Dos Santos CC, Deng Y, Lai PFH, Slutsky AS, et al. Mesenchymal stem cells reduce inflammation while enhancing bacterial clearance and improving survival in sepsis. American Journal of Respiratory and Critical Care Medicine. 2010;**182**:1047-1057

[70] Zhao X, Liu D, Gong W, Zhao G, Liu L, Yang L, et al. The toll-like receptor 3 ligand, poly(I:C), improves immunosuppressive function and therapeutic effect of mesenchymal stem cells on sepsis via inhibiting MiR-143. Stem Cells. 2014;**32**:521-533

[71] Mantovani A, Biswas SK, Galdiero MR, Sica A, Locati M. Macrophage plasticity and polarization in tissue repair and remodelling. The Journal of Pathology. 2013;**229**:176-185

[72] Martinez FO, Sica A, Mantovani A, Locati M. Macrophage activation and polarization. Frontiers in Bioscience: A Virtual Library of Medicine. 2008;**13**:453-461

[73] Krasnodembskaya A, Samarani G, Song Y, Zhuo H, Su X, Lee J-W, et al. Human mesenchymal stem cells reduce mortality and bacteremia in gram-negative sepsis in mice in part by enhancing the phagocytic activity of blood monocytes. American Journal of Physiology: Lung Cellular and Molecular Physiology. 2012;**302**:L1003-L1013

[74] Devaney J, Horie S, Masterson C, Elliman S, Barry F, O'Brien T, et al. Human mesenchymal stromal cells decrease the severity of acute lung injury induced by *E. coli* in the rat. Thorax. 2015;**70**:625-635

[75] Luz-Crawford P, Jorgensen C, Djouad F. Mesenchymal stem cells direct the immunological fate of macrophages. Results and Problems in Cell Differentiation. 2017;**62**:61-72

[76] Kim J, Hematti P. Mesenchymal stem cell-educated macrophages: A novel type of alternatively activated macrophages. Experimental Hematology. 2009;**37**:1445-1453

[77] Brekke O-L, Christiansen D, Fure H, Fung M, Mollnes TE. The role of complement C3 opsonization, C5a receptor, and CD14 in *E. coli*-induced up-regulation of granulocyte and monocyte CD11b/CD18 (CR3), phagocytosis, and oxidative burst in human whole blood. Journal of Leukocyte Biology. 2007;**81**:1404-1413

[78] Deng W, Han Q, Liao L, You S, Deng H, Zhao RCH. Effects of allogeneic bone marrow-derived mesenchymal stem cells on T and B lymphocytes from BXSB mice. DNA and Cell Biology. 2005;**24**:458-463

[79] Corcione A, Benvenuto F, Ferretti E, Giunti D, Cappiello V, Cazzanti F, et al. Human mesenchymal stem cells modulate B-cell functions. Blood. 2006;**107**:367-372

[80] Glennie S, Soeiro I, Dyson PJ, Lam EW-F, Dazzi F. Bone marrow mesenchymal stem cells induce division arrest anergy of activated T cells. Blood. 2005;**105**:2821-2827

[81] Najar M, Raicevic G, Fayyad-Kazan H, De Bruyn C, Bron D, Toungouz M, et al. Impact of different mesenchymal stromal cell types on T-cell activation, proliferation and migration. International Immunopharmacology. 2013;**15**:693-702

[82] Najar M, Raicevic G, Boufker HI, Fayyad Kazan H, De Bruyn C, Meuleman N, et al. Mesenchymal stromal cells use PGE2 to modulate activation and proliferation of lymphocyte subsets: Combined comparison of adipose tissue, Wharton's jelly and bone marrow sources. Cellular Immunology. 2010;**264**:171-179

[83] Djouad F, Plence P, Bony C, Tropel P, Apparailly F, Sany J, et al. Immunosuppressive effect of mesenchymal stem cells favors tumor growth in allogeneic animals. Blood. 2003;**102**:3837-3844

[84] Maccario R, Podestà M, Moretta A, Cometa A, Comoli P, Montagna D, et al. Interaction of human mesenchymal stem cells with cells involved in alloantigen-specific immune response favors the differentiation of CD4+ T-cell subsets expressing a regulatory/suppressive phenotype. Haematologica. 2005;**90**:516-525

[85] Najar M, Raicevic G, Fayyad-Kazan H, De Bruyn C, Bron D, Toungouz M, et al. Bone marrow mesenchymal stromal cells induce proliferative, cytokinic and molecular changes during the T cell response: The importance of the IL-10/CD210 axis. Stem Cell Reviews. 2015;**11**:442-452

[86] Nauta AJ, Fibbe WE. Immunomodulatory properties of mesenchymal stromal cells. Blood. 2007;**110**:3499-3506

[87] Mohammadpour H, Pourfathollah AA, Zarif MN, Tahoori MT. TNF-α modulates the immunosuppressive effects of MSCs on dendritic cells and T cells. International Immunopharmacology. 2015;**28**:1009-1017

[88] Alcayaga-Miranda F, Cuenca J, Martin A, Contreras L, Figueroa FE, Khoury M. Combination therapy of menstrual derived mesenchymal stem cells and antibiotics ameliorates survival in sepsis. Stem Cell Research & Therapy. 2015;**6**:199

[89] Gupta N, Krasnodembskaya A, Kapetanaki M, Mouded M, Tan X, Serikov V, et al. Mesenchymal stem cells enhance survival and bacterial clearance in murine Escherichia coli pneumonia. Thorax. 2012;**67**:533-539

[90] Krasnodembskaya A, Song Y, Fang X, Gupta N, Serikov V, Lee J-W, et al. Antibacterial effect of human mesenchymal stem cells is mediated in part from secretion of the antimicrobial peptide LL-37. Stem Cells. 2010;**28**:2229-2238

[91] Sung P-H, Chang C-L, Tsai T-H, Chang L-T, Leu S, Chen Y-L, et al. Apoptotic adipose-derived mesenchymal stem cell therapy protects against lung and kidney injury in sepsis syndrome caused by cecal ligation puncture in rats. Stem Cell Research & Therapy. 2013;**4**:155

[92] Pedrazza L, Lunardelli A, Luft C, Cruz CU, de Mesquita FC, Bitencourt S, et al. Mesenchymal stem cells decrease splenocytes apoptosis in a sepsis experimental model. Inflammation

Research: Official Journal of the European Histamine Research Society. 2014;**63**:719-728

[93] Bhattacharya J, Matthay MA. Regulation and repair of the alveolar-capillary barrier in acute lung injury. Annual Review of Physiology. 2013;**75**:593-615

[94] Pati S, Khakoo AY, Zhao J, Jimenez F, Gerber MH, Harting M, et al. Human mesenchymal stem cells inhibit vascular permeability by modulating vascular endothelial cadherin/β-catenin signaling. Stem Cells and Development. 2011;**20**:89-101

[95] Pati S, Gerber MH, Menge TD, Wataha KA, Zhao Y, Baumgartner JA, et al. Bone marrow derived mesenchymal stem cells inhibit inflammation and preserve vascular endothelial integrity in the lungs after hemorrhagic shock. PLoS One. 2011;**6**:e25171

[96] Lee JW, Fang X, Gupta N, Serikov V, Matthay MA. Allogeneic human mesenchymal stem cells for treatment of *E. coli* endotoxin-induced acute lung injury in the ex vivo perfused human lung. Proceedings of the National Academy of Sciences of the United States of America. 2009;**106**:16357-16362

[97] Xu Z, Zhang C, Cheng L, Hu M, Tao H, Song L. The microRNA miR-17 regulates lung FoxA1 expression during lipopolysaccharide-induced acute lung injury. Biochemical and Biophysical Research Communications. 2014;**445**:48-53

[98] Monsel A, Zhu Y, Gennai S, Hao Q, Hu S, Rouby J-J, et al. Therapeutic effects of human mesenchymal stem cell-derived microvesicles in severe pneumonia in mice. American Journal of Respiratory and Critical Care Medicine. 2015;**192**:324-336

[99] Wang X, Gu H, Qin D, Yang L, Huang W, Essandoh K, et al. Exosomal miR-223 contributes to mesenchymal stem cell-elicited cardioprotection in polymicrobial sepsis. Scientific Reports. 2015;**5**:13721

[100] Laffey JG, Matthay MA. Fifty years of research in ARDS. Cell-based therapy for acute respiratory distress syndrome. Biology and potential therapeutic value. American Journal of Respiratory and Critical Care Medicine. 2017;**196**:266-273

[101] Lalu MM, McIntyre L, Pugliese C, Fergusson D, Winston BW, Marshall JC, et al. Safety of cell therapy with mesenchymal stromal cells (SafeCell): A systematic review and meta-analysis of clinical trials. PLoS One. 2012;**7**:e47559

[102] Wilson JG, Liu KD, Zhuo H, Caballero L, McMillan M, Fang X, et al. Mesenchymal stem (stromal) cells for treatment of ARDS: A phase 1 clinical trial. The Lancet Respiratory Medicine. 2015;**3**:24-32

[103] McIntyre LA, Stewart DJ, Mei SHJ, Courtman D, Watpool I, Granton J, et al. Cellular immunotherapy for septic shock. A phase I clinical trial. American Journal of Respiratory and Critical Care Medicine. 2018;**197**:337-347

[104] Zheng G, Huang L, Tong H, Shu Q, Hu Y, Ge M, et al. Treatment of acute respiratory distress syndrome with allogeneic adipose-derived mesenchymal stem cells: A randomized, placebo-controlled pilot study. Respiratory Research. 2014;**15**:39

[105] Famous KR, Delucchi K, Ware LB, Kangelaris KN, Liu KD, Thompson BT, et al. Acute respiratory distress syndrome subphenotypes respond differently to randomized fluid management strategy. American Journal of Respiratory and Critical Care Medicine. 2017;**195**:331-338

[106] Calfee CS, Delucchi K, Parsons PE, Thompson BT, Ware LB, Matthay MA, et al. Subphenotypes in acute respiratory distress syndrome: Latent class analysis of data from two randomised controlled trials. The Lancet Respiratory Medicine. 2014;**2**:611-620

[107] Bustos ML, Huleihel L, Kapetanaki MG, Lino-Cardenas CL, Mroz L, Ellis BM, et al. Aging mesenchymal stem cells fail to protect because of impaired migration and antiinflammatory response. American Journal of Respiratory and Critical Care Medicine. 2014;**189**:787-798

[108] Matthay MA, Calfee CS, Zhuo H, Thompson BT, Wilson JG, Levitt JE, et al. Treatment with allogeneic mesenchymal stromal cells for moderate to severe acute respiratory distress syndrome (START study): A randomised phase 2a safety trial. The Lancet Respiratory Medicine. Feb 2019;7(2):154-162

[109] Galstyan GM, Makarova PM, Parovichnikova EN. Use of mesenchymal stromal stem cells for the treatment of sepsis. Anesteziologiia i Reanimatologiia. 2015;**60**:59-65

Section 2

Flow Cytometry

Chapter 2

Flow Cytometry Applied to the Diagnosis of Primary Immunodeficiencies

Mónica Martínez-Gallo and Marina García-Prat

Abstract

Primary immunodeficiencies are the result of biological defects associated with functional immune abnormalities. It consists of a group of disorders showing a higher incidence and severity of infections, expression of immunological dysregulation such as inflammation and lymphoproliferation. The immunophenotyping and *in vitro* functional characterization of immunodeficient patients contribute, together with the clinical aspects, to define the underlying immune defect particularities. Flow cytometry applications in primary immunodeficiency assessment are multiple and include the study of a wide range of specific cell lymphocyte subpopulations. This chapter describes the main techniques used in the diagnosis of a wide variety of primary immunodeficiencies, in which intracellular proteins or activation markers involved in immunity are evaluated, as well as functional proliferation, cytokine production, phosphorylation of transcription factors, cytotoxic and degranulation capacity. Flow cytometry is a tool that allows rapid and accurate evaluation of multiple lymphocyte populations and immunological function, and this information is essential for the diagnosis and evaluation of patients with primary immunodeficiencies.

Keywords: primary immunodeficiencies, lymphocyte subsets, extended immunophenotyping, flow cytometry, functional assays, intracellular cytokine production, cytotoxicity, phospho-stat flow

1. Introduction

Primary immunodeficiencies (PID) are a heterogeneous group of genetic disorders which manifest clinically as recurrent infections, autoimmune and/or autoinflammatory diseases or malignancies. In PIDs, the immune system is affected quantitatively or qualitatively and includes more than 400 different entities, with an incidence of approximately 1:2000 live newborns. The Expert Committee of the International Union of Immunological Societies has classified PIDs into nine groups based on the clinical manifestations and laboratory immunological abnormalities [1]. Prompt identification of PID patients reduces complications and is associated with a more favorable prognosis [2]. In addition to a complete medical history, physical examination, and general laboratory tests, the initial evaluation protocol when PID is suspected includes analysis of serum immunoglobulin levels and extended immunophenotyping in peripheral blood. Immunophenotype abnormalities can range from a complete absence of a specific cell population to more subtle variations in the differentiated states of specific subpopulations [3]. For basic and advanced lymphocyte phenotype studies, the technical option is classical flow cytometry, although new approaches mainly used in research like mass cytometry or CyTOF, have emerged [4]. The following step in the diagnosis of patients with PID is the biochemical and functional characterization of the altered molecules, for example, perforin, CD40 ligand, etc. The advantage of using flow cytometry over other techniques such as western-blot in the case of protein expression studies and functional evaluation of different subpopulations is the rapid and sensitive result.

2. Evaluation of immunophenotype to aid in the differential diagnosis of PIDs

Flow cytometry is a standard laboratory tool in the evaluation and identification of leukocyte populations and specific lymphocyte subpopulations. The analysis of the expression of cell surface markers [clusters of differentiation (CD)] allows to determine the cell lineage and to examine the stage of differentiation and cell activation. The clinical application is broad for routine use in diagnostic laboratories, which facilitates the assessment of patients with suspected primary immunodeficiencies. The circulating cellular compartment of the human immune system shows the most marked shifts during the first years of life, but is also dynamic during adult life, showing a tendency to a reduction in function and diversity with age [5]. When an immune system disorder is suspected, immunophenotyping of the patient's peripheral blood mononuclear cells (PBMCs) is one of the initial steps in the diagnostic work-up [6].

The basic immunophenotype analyzing T (CD3), B (CD19), and natural killer (NK) (CD56+CD16+) lymphocytes leads to a potential diagnosis of severe combined immunodeficiencies (SCID) resulting in four general categories of disease defined by the impact on lymphocyte populations: T−/B−/NK−, T−/B−/NK+, T−/B+/NK−, and T−/B+/NK+. Overall, patients with adenosine deaminase (ADA) deficiency and reticular dysgenesis demonstrate the T−/B−/NK− immunophenotype and typically have the most profound lymphocytopenia among SCID patients [7]. Recombination defects usually show the phenotype T−/B−/NK+, and defects in cytokine signaling and T-cell receptor (TCR)-derived signals result in T−/B+/NK− and T−/B+/NK+ phenotypes. These phenotypes often differ depending on the type of mutation and protein affected. The expression of isoforms CD45RA, CD45RO in T cells should also be included in the study of patients with suspected SCID. An increase in the percentage of CD45RO memory helper T cells indicates the presence of autoreactive lymphocytes and reinforces the diagnostic suspicion of SCID, leaky SCID or Omenn syndrome (OS) (**Figure 1**). OS is an inherited disorder characterized by an absence of circulating B lymphocytes and an infiltration of the skin by activated T lymphocytes. Inherited mutations in RAG1 or RAG2, resulting in partial V(D)J recombinase activity, were shown to be responsible for OS [8].

The use of larger cytometric panels can include a higher number of lymphocyte subpopulations of interest in the diagnosis of other groups of PID, such as predominantly antibody deficiencies and immune dysregulation diseases among others. Below we describe the gating strategies for the study of a broader spectrum of subpopulations: T cell differentiation stages, recent thymic emigrants (RET), regulatory T cells (Tregs), T helper 1/2/17 cells, B cell differentiation stages, NK cells, dendritic cells and monocytes, and their application in the study of patients with PID.

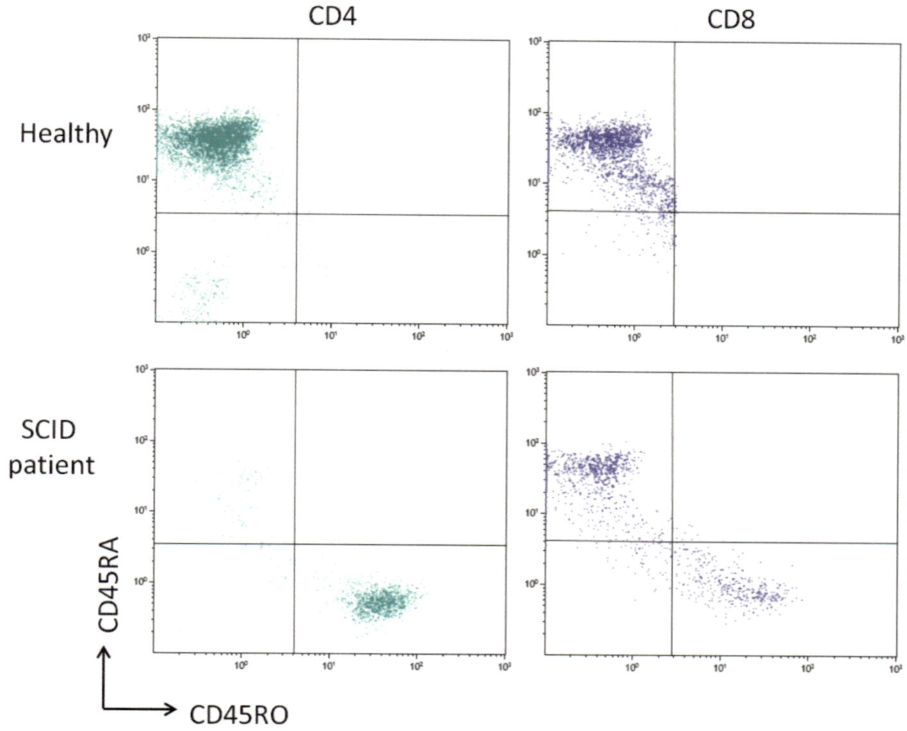

Figure 1.
Dotplot representing the expression of different isoforms CD45RA and CD45RO in subpopulations T CD4 and CD8. A healthy donor and a patient with severe combined immunodeficiency (SCID) of the same age range are compared.

General immunophenotyping protocol:

1. Stain 50 μL of EDTA whole blood with appropriate mix of antibodies of interest, depending on the panel.

2. Incubate for 20 min in the dark at room temperature (RT).

3. Lyse the red cells using a lysing solution.

4. Incubate for 10 min in the dark at RT.

5. Discard the lysing solution after 5 min of centrifugation at (324 g) or 1500 revolutions per minute (rpm).

6. Wash with 2 mL of phosphate buffered saline (PBS).

7. Discard the PBS after 5 min of centrifugation at 1500 rpm.

8. Add 200 μL of PBS, vortex and the sample is ready to be acquired by flow cytometry.

(This protocol is used in all the immunophenotyping techniques, using the appropriate mix of antibodies in each situation).

2.1 T lymphocyte immunophenotyping

T-lymphocytes mature in the thymus after going through positive and negative selection processes before migration into secondary lymphoid organs. The immune system at birth is largely dominated by truly naïve T cells, with very limited representation of memory cells and a relatively high proportion of recent thymic emigrants (RTEs). They are naïve peripheral T cells, which have only recently exited the thymus, have not undergone further peripheral proliferation and exhibit phenotypic and functional characteristics distinct from those of their more mature counterparts in the naïve peripheral T cell pool [9]. Thymic function can be determined by evaluation RTEs, which are studied using CD3, CD4, CD27, CD31, and CD45RA, expression markers according to the gating strategy in (**Figure 2**). The assessment of the thymic function is performed also in a suspicion of severe combined immunodeficiency, DiGeorge syndrome, and other T-cell immune deficiency disorders.

Subsequently, the mature "naïve" T-lymphocytes circulate between the secondary lymphoid organs and the blood. In humans, naïve CD4+ T cells typically express CCR7, CD62L, and CD45RA, while they lack CD45RO expression [10]. CCR7 and CD62L are involved in the localization of T cells in secondary lymphoid organs (SLO) and interact with ligands expressed in high endothelial venules (HEVs). CD45RA and CD45RO play a role in the transduction of TCR signals, and their expression characterizes the different subsets of T cells, being CD45RA positive the naïve and CD45RO positive the memory [11]. Upon recognition of specific antigen complexes, naive CD4+ T cells proliferate and differentiate toward effector T cells, which provide immediate protection.

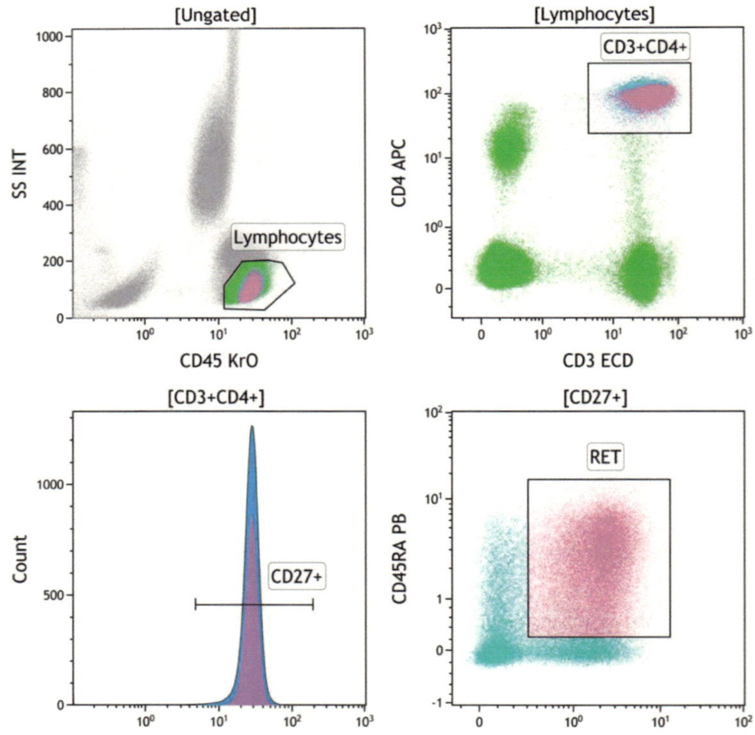

Figure 2.
Gating strategy of recent thymic emigrants (RET). RTEs were studied using CD3, CD4, CD27, CD31, and CD45RA expression.

Most of these effectors T cells subsequently die from apoptosis, but a subset of antigen-specific T cells will persist as memory T cells [12]. There are two types of memory T cells in circulation, central memory T cells (TCMs) and effectors memory T cells (TEM): the former shows self-renewal potential with less effector functions, while the latter possesses immediate effector functions and can migrate rapidly to peripheral tissues. In addition, a distinct lineage of tissue-resident memory cells (TEMRA cells) is limited to different tissues. By using the characteristic expression markers CD45RA and CCR7 and a gating strategy for gated CD4+ and CD8+ T-cell subsets, we can identify the following populations CD45RA+/CCR7+ (naïve), CD45RA−/CCR7+ [central memory (TCM)], CD45RA−/CCR7− [effector memory (TEM)], and CD45RA+/CCR7− [terminal effector memory (TEMRA)]. Furthermore, based on CCR6 and CXCR3 expression markers, we can distinguish CD4+ T-helper (Th) populations at phenotypic level (Th1, Th2, Th17, and Th1-17) by gating on CD4+CD45RA− TCM and TEM cells (**Figure 3**).

In several primary immunodeficiencies, there are alterations in different subpopulations. The distribution of naïve and memory T cells is particularly relevant for the diagnosis of severe combined immunodeficiency (SCID) and its variants in infants—Omenn syndrome and leaky SCID.

In patients with antibody defects such as common variable immunodeficiency (CVID) has been described an increase abnormally differentiated T CD4+ effector cells showing a Th1-skewed profile and activated effector T CD8+ lymphocytes [13, 14].

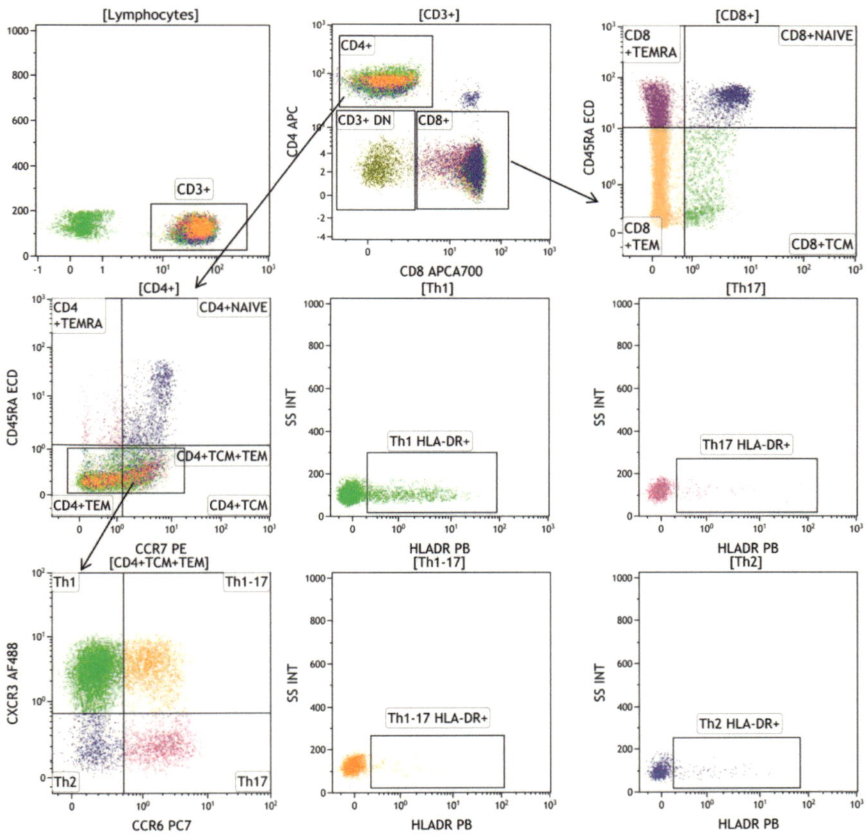

Figure 3.
Gating strategy for differentiated CD4+ and CD8+ T cell subsets, based on CD45RA and CCR7 expression defining: CD45RA+/CCR7+ (naïve), CD45RA−/CCR7+ (central memory [TCM]), CD45RA−/CCR7− (effector memory [TEM]), and CD45RA+/CCR7− (terminal effector memory [TEMRA]). CD4+ T-helper (Th) populations (Th1, Th2, Th17, and Th1-17), based on CCR6 and CXCR3 expression, were analyzed by gating on CD45RA− TCM and TEM cells.

Also with the suspected diagnosis of hyper-IgE syndrome (atopic dermatitis, recurrent otitis, pneumonias, and abscess caused by *Staphylococcus aureus*), the use of this gating strategy is very useful. The immunological phenotypes include high IgE levels, eosinophilia, low levels of Th17 cells, and memory B cell lymphopenia. T-cell differentiation, particularly Th17 cells is disrupted, for that reason, we use a flow cytometry panel to analyze the CD4+ T-helper (Th) populations (Th1, Th2, Th17, and Th1-17), based on CCR6 and CXCR3 expression, previously gated on CD45RA− TCM and TEM CD4+ T cells. It is important to compare the values of the patient with an age-matched healthy control [6].

Regulatory T-cells (Tregs) play a key role in maintaining peripheral tolerance under physiological and pathological conditions. Tregs are characterized by the constitutive expression of the critical transcription factor forkhead box P3 (Foxp3). Impairment of Treg function is the key pathogenic event leading to altered self-tolerance in patients with immune deregulation, poly-endocrinopathy, enteropathy, and X-linked syndrome (IPEX) [15]. The most common markers utilized for Treg typing by flow cytometry are CD25, forkhead box protein 3 (FoxP3), and CD127. Also for more accurate characterization, CCR4 is useful; a potential marker for effector-type regulatory T cells T-regulatory (Treg) cell populations: CD3+CD4+CD25+, CD127−, CCR4+, and CD45RO+.

See the immunophenotype protocol in Section 2.

2.2 B lymphocyte immunophenotyping

Early development of human B cells from multipotent stem cells occurs in the bone marrow by antigen independent manner. Once the central development of B cells has been accomplished and the transitional B cells leave the bone marrow, maturation occurs in the secondary lymphoid organs, especially in the spleen.

B-cell subset phenotyping in peripheral blood has been used to classify and assess the prognosis of patients with humoral deficiencies such as common variable immunodeficiency (CVID). CVID is a heterogeneous disease, and most patients show defective differentiation of B cells into memory B cells or antibody-secreting cells. B cells can be identified in the lymphocyte gate by CD19 expression, a marker that is present in all circulating B cell populations. B-cell populations can be studied based on expression of IgD and CD27 [naïve (IgD+CD27−), pre-switched (IgD+CD27+), switched memory (IgD−CD27+), and exhausted (IgD−CD27−)]. In patients with CVID, the study of these subpopulations allows to identify severe defects of germinal center dependent B cell memory formation and early defects of peripheral B cell differentiation. The absence of switched memory B cells is very suggestive of an alteration of the germinal center reaction [16].

The combined staining for CD19, IgD, CD38, and CD24 allows the identification of transitional B cells (CD19+IgD+CD38^{high}CD24^{high}) and plasmablasts (CD19+IgD+CD38^{high}CD24^{neg}). CD27 and CD21 enabled study of the CD21^{low} population. The expansion of circulating CD21^{low}B cells is associated with manifestations of a chronic immune activation and autoimmunity (**Figure 4**) [17].

Recently, germline heterozygous gain-of-function mutations in PIK3R1 and PIK3CD genes have been identified to cause a primary immunodeficiency in patients with node lymphoproliferation, splenomegaly, diarrhea, and dysgammaglobulinemia, initially diagnosed with CVID. In these patients, the extended immunophenotyping reveals high percentage of transitional B cells, low pre-switched memory, switched memory B cells, and skewing of CD4+ with CD8+ T cells toward terminally differentiated effector cells. This phenotype is suggestive of activated PI3K delta syndrome type 1 and 2 (APDS) [18].

See the immunophenotype protocol in Section 2.

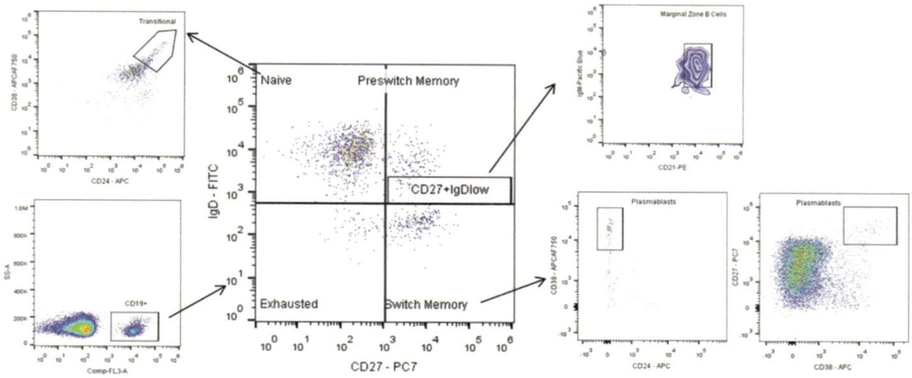

Figure 4.
Gating strategy used for the study of subpopulations of B lymphocytes. B-cell populations (naïve, pre-switched, switched memory, and exhausted) are distinguished depending on IgD and CD27 expression. The differing pattern of CD24+ and CD38+ expression identified transitional cells and plasmablasts. CD27 and CD21 markers enabled study of the CD21 low population.

2.3 Innate cells immunophenotyping

2.3.1 Natural killer (NK) cells

Natural killer (NK) cells are one of the first lines of innate immune defense against pathogens and tumors, among other cells. NK cells are lymphocytes with cytotoxic capacity that provide rapid response to virus-infected cells and to tumor cells through the ability to recognize the absence of HLA class I and other signals on these cells. Innate cells can be analyzed by flow cytometry gating in the $CD3^{-}CD19^{-}$ cells. Natural killer (NK) cells subpopulations can be divided into two subpopulations depending on CD56 and CD16 expression; NK^{dim} ($CD16^{+}CD56^{low}$) and NK^{bright} ($CD16^{-}CD56^{high}$). Characterizing the different subsets of NK is crucial in the diagnosis of NK cell deficiencies [19]. An example is the deficiency in MCM4, with autosomal recessive inheritance (AR), which may be suspected in the context of a higher percentage of CD56bright NK cells (of total NK cells) in individuals with endocrine and/or growth abnormalities [20].

See the immunophenotype protocol in Section 2.

2.3.2 Monocytes and dendritic cells (DC)

Monocytes are leukocytes that can differentiate into macrophages and myeloid lineage dendritic cells. As a part of the vertebrate innate immune system, monocytes also influence the process of adaptive immunity [21]. There are at least two subpopulations of peripheral monocytes in human blood based on their phenotypic receptors CD16 and CD14; classical monocytes ($CD14^{high}CD16^{-}$) and non-classical monocytes ($CD14^{low}CD16^{+}$). These cells have been shown to exhibit distinct phenotype and function [22].

Dendritic cells in periphery blood are responsible for the capture and processing antigens, express lymphocyte co-stimulatory molecules, migrate to lymphoid organs, and secrete cytokines to initiate immune responses [23]. Dendritic cells were studied into the population negative for the following markers: CD3, CD14, CD16, CD19, CD20, and CD56. High expression of HLA-DR, CD11c, and CD123 are markers commonly used to identify plasmacytoid DCs ($HLA\text{-}DR^{+}CD123^{+}$) and myeloid DCs ($HLA\text{-}DR^{+}CD11c^{+}$) (**Figure 5**). GATA binding protein 2 (GATA-2) deficiency is an autosomal dominant (AD) immunodeficiency classified in the

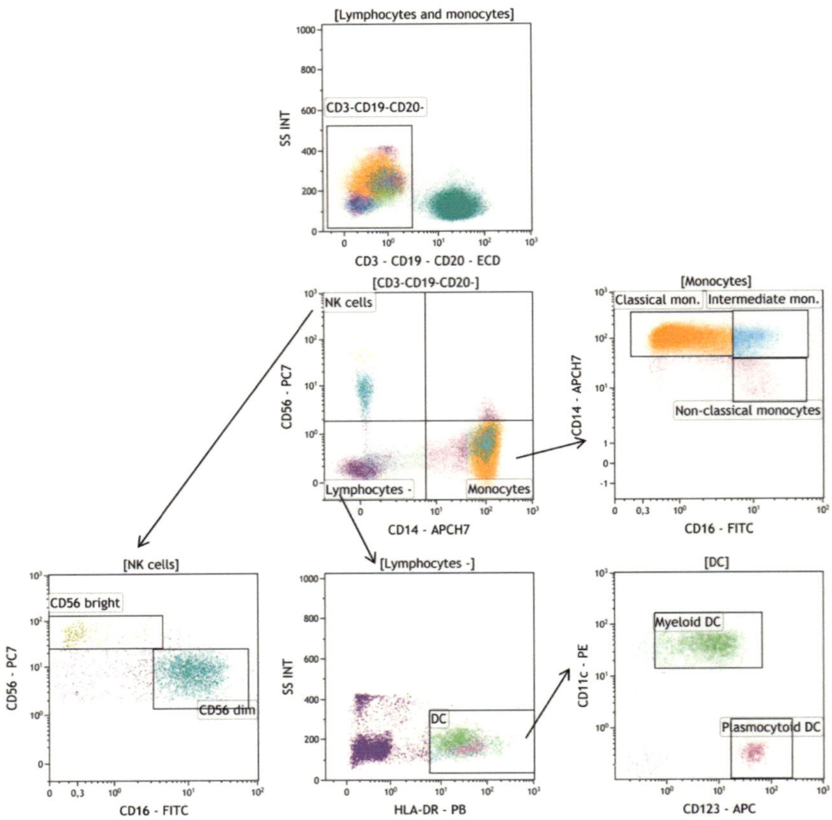

Figure 5.
NK/monocytes/DC dendritic cells (DC) gating strategy. NK cells and monocyte populations were analyzed in the CD3−CD19− gate. NK subpopulations (NK^{dim} and NK^{bright}) were studied using CD56 and CD16 expression markers. The markers CD16 and CD14 were used to identify classical monocytes (CD14+CD16−) and non-classical monocytes (CD16+CD14−). DCs were studied selecting the population negative for the following markers: CD3, CD14, CD16, CD19, CD20, and CD56. High expression of HLA-DR and CD11c and CD123 was used to identify plasmacytoid DCs (HLA-DR+CD123+) and myeloid DCs (HLA-DR+CD11c+).

group of congenital defects of the number or function of phagocytes in which the following leukocyte populations are affected: $CD56^{bright}$ NK cell, B lymphocytes, monocytes and peripheral dendritic cells populations are affected. Patients with this PID have susceptibility to infections by mycobacteria and human papillomavirus (HPV), histoplasmosis, alveolar proteinosis, myelodysplastic syndromes such acute myelogenous leukemia and lymphedema [24].

See the immunophenotype protocol in Section 2.

3. Evaluation of intracellular proteins and activation markers involved in immunity

3.1 Evaluation of intracellular proteins

Intracellular protein and specific cell surface antigen detection is a screening test for PIDs. Depending on the specific clinical presentation of the patient, the diagnostic suspicion can be assessed by evaluating the expression of the affected protein by flow cytometry. The most widely used procedure for staining intracellular molecules requires the fixation of cells in suspension and subsequent permeabilization before adding the detection antibody. This fixation/permeabilization treatment allows the antibody to pass through the plasma membrane into the cell while maintaining morphological characteristics. While this technique is informative and very useful to identify total defects of a particular protein, detecting expression does not rule out the defect, as there may be mutations that do not alter the expression but the function. Some examples of this type of application to PID diagnosis are the evaluation of the following proteins:

For intracellular staining, there are several commercial options, including laboratory permeability and fixation reagents. It is important to take into account that not all commercial kits work the same way for different cell types or for different techniques. It is therefore necessary to adjust the conditions in each particular case.

Intracellular staining protocol:

1. Fix 50 μL of EDTA whole blood with the appropriate volume of fixative reagent.

2. Incubate for 15 min in the dark at RT.

3. Add the appropriate volume of permeabilizing reagent.

4. Add the extracellular antibodies mix.

5. Add the appropriate intracellular antibody.

6. Incubate for 20 min at RT and in the dark.

7. Add the appropriate volume of wash buffer.

8. Discard the wash buffer after 5 min of centrifugation at 1500 rpm.

9. Add 100 μL of PBS and the sample is ready to be acquired by flow cytometry.

(This protocol may change slightly depending on the commercial kit used).

3.1.1 BTK expression in X-linked agammaglobulinemia

X-linked agammaglobulinemia or Bruton tyrosine kinase (BTK) deficiency is a disorder caused for mutations in the gene encoding BTK. It is a tyrosine-protein kinase localized intracellularly next to the B-cell receptor (BCR), and it is crucial for the B cell development. The expression of BTK protein is evaluated in monocytes or platelets as patients with suspected X-linked agammaglobulinemia have an absent or significantly reduced number of circulating B lymphocytes. This leads to the fact that intracellular BTK protein has to be studied by gating monocytes [25].

See the intracellular staining protocol in Section 3.1.

3.1.2 SAP, XIAP, and perforin expression in Hemophagocytic lymphohistiocytosis (HLH)

Measurement of intracellular signaling lymphocyte activation molecule-associated protein and X-linked inhibitor of apoptosis are used as screening tools to evaluate X-linked lymphoproliferative syndrome disorder type 1 and 2, respectively. This approach has been shown to provide good sensitivity and specificity compared with genetic mutation evaluation, and it is characterized by high negative predictive value. The lack of intracellular perforin expression in NK cells is characteristic of familial hemophagocytic lymphohistiocytosis type 2 [26, 27].

See the intracellular staining protocol in Section 3.1.

3.1.3 FOXP3 expression in IPEX (immunodysregulation polyendocrinopathy enteropathy X-linked) syndrome

A further example of intracellular protein testing by flow cytometry in patients with PIDs involves screening for patients with possible immune dysregulation, polyendocrinopathy, enteropathy, and X-linked syndrome (the gene encoding FOXP3 protein is located on the X chromosome, therefore only males are affected by the disease while females are carriers). This diagnosis is confirmed in male patients whose CD4+CD25+ regulatory T cells demonstrate absence or decreased expression of FOXP3 [28].

See the intracellular staining protocol in Section 3.1.

3.2 Immediate early activation markers

The activation of T lymphocytes is a complex and finely regulated process of events resulting in the expression of cytokine receptors, the production and secretion of cytokines and increased cell surface molecules that facilitate the immune response. The most important activation molecules expressed on T lymphocytes can be classified as early activation markers, such as CD69 and CD25, and late activation markers, such as CD62L and HLA-DR. Additionally, very late activation markers, such as VLA-1 have been described, playing a role in lymphocyte adhesion and extravasation [29].

CD69 is a signaling membrane glycoprotein involved in the induction of T cell proliferation. CD69 is expressed at very low levels T cells at rest in PBMCs (<5–10%) and is one of the first evaluable activation markers, being rapidly increased in T cells within the first hour of TCR. Expression of CD69 peaks at 16–24 h and then decreases. The inability to upregulate CD69 following TCR activation may be associated with T cell dysfunction [30].

CD40 ligand (CD40L), also called CD154, is a member of the TNF-receptor superfamily, primarily expressed on activated T cells, and functions as a co-stimulatory molecule by binding CD40 which is constitutively expressed on antigen presenting cells (APCs). The CD40L-CD40 ligation is a fundamental interaction for B lymphocytes proliferation, formation of germinal centers, Ig class switch, antibodies maturation, and B lymphocyte differentiation to plasmatic cell and to memory B lymphocyte. Deficiency of CD154 is an X-linked disorder that only males are affected and females are the carriers, also known as Hyper IgM syndrome [31]. Expression of CD40L on resting T cells from healthy donors is very low (<1%) and is quickly upregulated within 1–2 h after TCR stimulation via the transcription factors NFAT and AP-1. CD40L expression peaks near 6 h after stimulation and declines by 16–24 h (**Figure 6**) [32].

In summary, CD69 and CD40L are both rapidly induced following T cell activation and both exert important functions in T cell biology. Expressions of these markers have both been shown to be altered in patients with different PIDs.

Immediate early activation markers staining protocol:

1. Stimulate heparin whole blood or peripheral blood mononuclear cells (PBMCs) with the appropriate stimuli or mitogen.

2. Incubate the appropriate time depending on the activation marker of your interest.

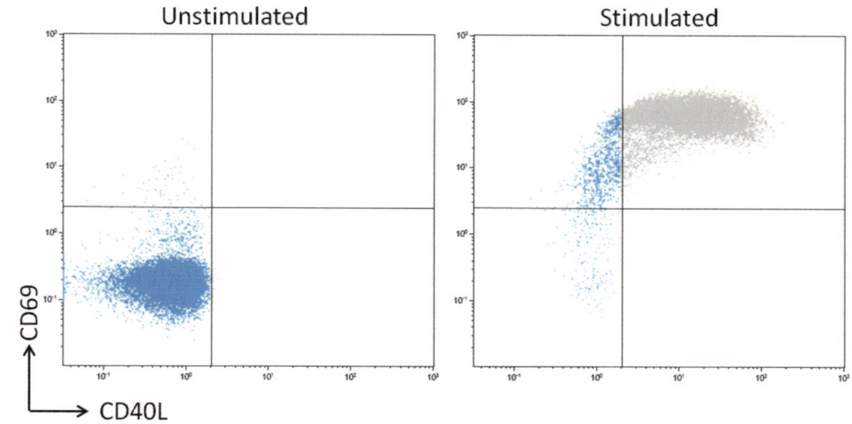

Figure 6.
The expression of CD40 ligand is studied by stimulating PBMCs with PMA plus ionomycin for 4 h, and then the expression of a cell activation marker such as CD69 is analyzed. Dotplots of PMBCs without stimulation (left) and stimulated (right) where a normal expression of CD40L and CD69 is observed.

3. Stain with appropriate mix of antibodies of interest.

4. Incubate for 20 min at RT and in the dark.

5. Lyse the red cells using 2 mL of lysing solution.

6. Incubate for 10 min at RT and in the dark.

7. Discard the lysing solution after 5 min of centrifugation at 1500 rpm.

8. Wash with 2 mL of PBS.

9. Discard the PBS after 5 min of centrifugation at 1500 rpm.

10. Add 100 μL of PBS and the sample is ready to be acquired by flow cytometry.

(The same protocol is useful for all the different activation marker techniques, changing mitogens, and incubation time depending on each case).

4. Evaluation of cellular function

The assessment of lymphocyte function typically refers to the study of T cell function through cell proliferation in response to polyclonal nonspecific and specific stimuli; however, this concept is more extensive and refers to any method that evaluates any of the various effectors or regulatory aspects of subset of lymphocytes [33]. As we have seen previously, these studies are of particular importance when the functional impact of a mutation is unknown; in such cases, flow cytometry provides a number of different immune function assessment techniques, including evaluation of specific immune cell line functions, cell activation, proliferation, and cytokine production.

4.1 Lymphocyte proliferation assay

Lymphocyte phenotype studies should be accompanied by laboratory tests assessing the functional immunity of T cells when a primary immunodeficiency is suspected [34]. Lymphocyte proliferation assay is used to determine lymphocyte activation and their ability of carry out cell-mediated immune response. Ligand binding and transmembrane signal transduction result in cell activation, proliferation of B and T cells which ultimately leads to clonal expansion. Flow cytometry techniques are used to evaluate *in vitro* different cell types proliferation, including the use of fluorescent follow-up cell tracking intracellular dyes [e.g., succinimidyl ester of carboxyfluorescein diacetate (CFSE)] which are capable of covalent binding to cytoplasmic structures of the cell and remain there for a long period of time, being redistributed into daughter cells by each cell division resulting in a 50% decrease in fluorescence intensity after activation for each cell division cycle [35]. Almost all lymphocytes can be stimulated to proliferate non-specifically by stimulating them *in vitro* with polyclonal or mitogenic stimulants such as phytohemagglutinin (PHA), concanavaline A, mitogen pokeweed (PWM), or anti-CD3 antibodies alone or in combination with anti-CD28 or IL-2. Generally, these trials involve at least 72-h or 6–7 days of culture depending on the stimulation (**Figure 7**). The use of flow cytometry has gradually replaced "gold standard" methods such as lymphocyte proliferation by incorporating [3H] thymidine into the DNA of dividing cells and measured using a liquid scintillation counter, which is proportional to the amount of cell proliferation. In terms of complexity, methods based on flow cytometry are considerably simpler to perform, and the undoubted advantage of this method is that the cells grow without any negative influence during the entire culture period, compared to thymidine. However, the interpretation is less objective and time-consuming, thus requiring experienced staff [36].

Cell proliferation induced by an activation signal is a critical parameter in the diagnosis of patients with a possible T-cell defect, including infants with abnormal T-cell receptor excision circles (TRECs) during newborn screening test. In these cases, the T-cell response to polyclonal stimuli is usually less than 10% of the lower limit of the reference value. However, there are patients with hypomorphic mutations that allow some degree of T cell response, often <30% of the lower limit of the reference value. Some laboratories use more than one dose of mitogen or mitogens, although this is not done routinely and is generally not necessary to detect significant defects in T cell immunity. There have been rare cases where mitogen-induced

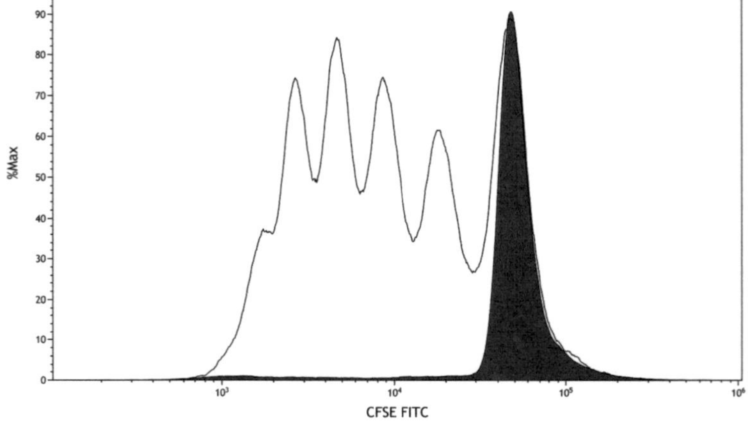

Figure 7.
Histograms of lymphocyte proliferation analysis after stimulation with PMA plus ionomicin. Lymphocytes staining with CFSE lose fluorescence after cell division. The histogram on the right shows the result of a patient, with no cell division shown compared to the result of a healthy donor on the histogram on the left.

proliferation was found to be abnormal, but further evaluation with a combination of agents that directly activate T cells [forbol 12-miristate 13-acetate (PMA) and ionomycin] showed that the proliferative capacity of T cells was normal, but the signaling apparatus (i.e., T[TCR]-CD3 cell antigen receptor-complex) was dysfunctional [7]. Obviously, one of the advantages of using cytometry to evaluate lymphocyte proliferation is the elimination of radioactive reagents. It also allows distinguishing proliferating cell subpopulations, but evaluation of the results is more subjective.

Lymphocyte proliferation assay protocol (CFSE):

1. Add 1 μL of carboxyfluorescein succinimidyl ester (CFSE) in 1 mL of PBMCs (all the PBMCs of your sample).

2. Manually gently stir the sample during 5 min.

3. Add 10 mL of Roswell Park Memorial Institute (RPMI) medium and centrifugate at 1500 rpm (324 g) during 5 min.

4. Count and adjust cells at 10^6 cells/mL.

5. Acquire 50 μL of the sample by flow cytometry just to make sure your sample has been correctly stained with the CFSE.

6. Dispense 100 μL of the PBMCs at 10^6 cells/mL in a 96 well plate.

7. Leave one well unstimulated (with RPMI) and add 100 μL of mitogens of interest in each different well [phorbol myristate acetate (PMA), ionomycin, phytohaemagglutinin (PHA), muromonab-CD3 (OKT3), interleukin-2 (IL-2), pokeweed mitogen (PWM), and concanavalin A (ConA)].

8. Incubate during 6 days at 37°C and in the dark.

9. Resuspend cell button and transfer it to fluorescence-activated cell sorter (FACS) tubes.

10. Stain with appropriate mix of antibodies of interest.

11. Incubate for 20 min at room temperature (RT) and in the dark.

12. Wash with 2 mL of phosphate buffered saline (PBS).

13. Discard the PBS after 5 min of centrifugation at 1500 rpm.

14. Add 100 μL of PBS and the sample is ready to be acquired by flow cytometry.

4.2 Functional analysis studying STAT phosphorylation

Signal transducer and activator of transcription (STAT) control cellular activation processes after cell activation through cytokines. All STAT molecules are phosphorylated by receptor-associated kinases, that causes activation, by forming homo- or heterodimers and finally translocate to nucleus to function as transcription factors [37]. The study of the different STATs by flow cytometry is of great interest not only of STAT deficiencies (STAT1, STAT3 or STAT5 deficiencies) but also to study the proper functioning of a signaling pathway in particular. When studying STAT phosphorylation, three essential factors must be taken into account: (A) the cell lineage in which STAT phosphorylation will be studied. In the study of PID patients, it is a limitation since it is usual that not all cell lineages are present; (B) choosing the correct recombinant cytokine is crucial as cytokine receptors are expressed differently in distinct lymphocyte lineages; (C) and selecting the appropriate antibodies against the STAT phosphorylation sites [30]. These studies can also be performed by WB, which is a very sensitive but more time-consuming technique and delays results [38].

STAT phosphorylation staining protocol:

1. Stain 100 μL of heparin whole blood with appropriate mix of antibodies of interest.

2. Incubate for 20 min at RT and in the dark.

3. Wash with 2 mL of PBS.

4. Discard the PBS after 5 min of centrifugation at 1500 rpm.

5. Add 100 μL of appropriate stimuli.

6. Incubate for 20 min at 37°C and in the dark.

7. Lyse the red cells using 1 mL of lyse and fix solution.

8. Incubate for 15 min at 37°C and in the dark.

9. Discard the lyse and fix solution after 5 min of centrifugation at 1500 revolutions per minute (rpm).

10. Wash with 1 mL of PBS with fetal bovine serum (FBS) at 10%.

11. Discard the PBS + FBS 10% after 5 min of centrifugation at 1500 rpm.

12. Add 500 μL of permeabilization buffer and incubate 30 min on ice or 4°C and in the dark.

13. Add the appropriate volume of intracellular anti-phosphorylated-STAT antibody.

14. Incubate for 30 min at RT and in the dark.

15. Wash with 1 mL of PBS with FBS at 10%.

16. Discard the PBS + FBS 10% after 5 min of centrifugation at 1500 rpm.

17. Add 100 μL of PBS + FBS 10% and the sample is ready to be acquired by flow cytometry.

(The same protocol is useful for the assay of different STATs. Mitogens/cytokines and antibodies may change depending on the STAT you are studying. Note that different commercial kits can be used to assess phospho-proteins, so the protocol can be slightly different in each case).

4.2.1 STAT1 phosphorylation for the diagnosis of chronic mucocutaneous candidiasis

STAT1 is a transcription factor which in humans can be activated by several ligands such as interferon alpha (IFNα), interferon gamma (IFNγ), epidermal growth factor (EGF), platelet derived growth factor (PDGF) or interleukin 6 (IL-6). Mutations in the STAT1 molecule can be gain of function (GOF) or loss of function (LOF). Both of them can cause PID with different phenotypes and symptoms [39].

STAT1 GOF dominant mutations were first discovered in patients with chronic mucocutaneous candidiasis (CMC). This disease typically shows symptoms such as persistent infections of the skin, oral/genital mucosa and nails caused mainly by *Candida albicans*. CMC may very often result from primary immunodeficiency. To assess the presence of a STAT1 GOF defect, isolated PBMCs are stimulated *in vitro* with INFγ. In order to know if there is a gain of function, we perform different experimental conditions, such as unstimulated, stimulated with INFγ, and with INFγ plus *Staurosporine*. *Staurosporine* is an inhibitor for $Ca^{2+/}$calmodulin-dependent protein kinase II inhibiting STAT1 phosphorylation [40]. The use of *Staurosporine* is justified to clarify the difference between control subjects (STAT1 WT alleles) and the CMC patients (STAT1 GOF alleles). STAT1 phosphorylation is assessed in CD14+ monocytes [41]. In a healthy donor, we can observe that STAT1 phosphorylation is reduced with the addition of Staurosporine, while in a patient with STAT1 GOF, phosphorylation is maintained despite the use of Staurosporine (**Figure 8**) [42].

Regarding **STAT1 LOF autosomic recessive mutations**, there are two genetic deficiencies that compromise the response to interferons type I and III. On the one hand, there is autosomal recessive deficiency (RA), partial or even complete STAT1. These types of defects are diagnosed in patients with an increase in intracellular or viral bacterial infections, with impaired responses to IFN α, β, γ, and IL27. In complete recessive form, there is a very low response to anti-viral and antimycotical treatments. On the other hand, partial STAT1 deficiency can also be autosomal dominant (AD); phenotypically causes impaired IFNγ responses and patients suffer with selective intracellular bacterial diseases such as Mendelian susceptibility to mycobacterial diseases (MSMD). To assess if there is a STAT1 LOF, isolated PBMCs

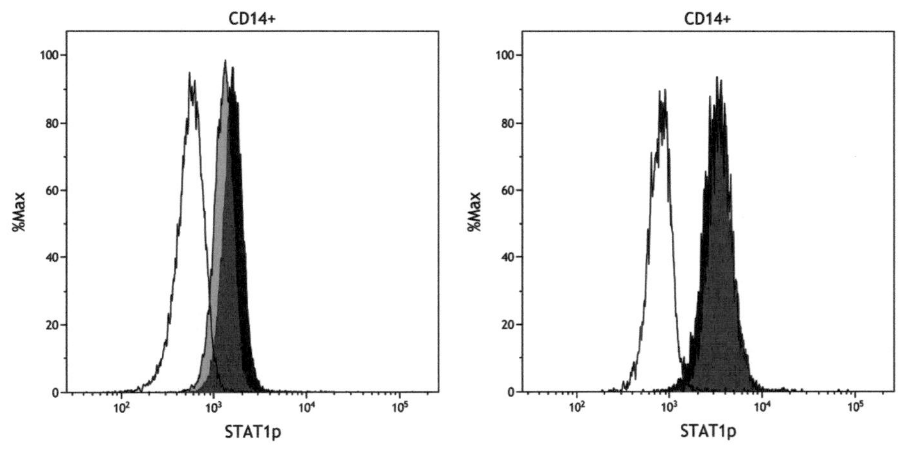

Figure 8.
The STAT1 phosphorylation and de-phosphorylation study for the diagnosis of function gain mutations (GOF) is evaluated in CD14+ monocytes. In a healthy donor (left), we can see that STAT1 phosphorylation is reduced with the addition of staurosporine (gray) while in the patient (right), phosphorylation is maintained despite the use of staurosporine (black: no staurosporine, white: control of isotypes).

are stimulated with INFγ and phosphorylation is assessed in CD14+ monocytes. In these cases, STAT1 phosphorylation in monocytes is clearly diminished comparing with a healthy donor [43].

See STAT phosphorylation staining protocol in Section 4.2.

4.2.2 STAT3 phosphorylation for the diagnosis of Hiper IgE Syndrome (HIES)

Hiper IgE Syndrome (HIES) is a PID defined as an association of atopy in a context of very high serum IgE levels, characteristic bacterial and fungal diseases, low-level clinical and biological inflammation, and various non-hematopoietic developmental manifestations. Somewhat arbitrarily, three disorders were successively put forward as the underlying cause of HIES: autosomal dominant (AD) STAT3 deficiency, the only disorder corresponding to the original definition of HIES, and autosomal recessive (AR) DOCK8 and PGM3 deficiencies, in which atopy and high serum IgE levels occur in a context of manifestations not seen in patients with typical HIES. Indeed, these three disorders disrupt different molecular pathways, affect different cell types, and underlie different clinical phenotypes [44].

Phenotype tests discussed in previous chapters, demonstrating low number of Th17 lymphocytes, may guide in-depth study of a defect in STAT3 signaling. These are the analysis of the ability to produce IL-17 by stimulation with PMA and ionomicin, adding a Golgi Plug such as Brefeldin and subsequent intracellular staining against IL17. And the STAT3 phosphorylation test in lymphocytes. Whole blood is stimulated with IL-6 and/or IL-21 and intracellular staining is performed to evaluate STAT3 phosphorylation. Different conditions are needed, such as: not stimulated, stimulated, and both with STAT3 antibodies or with an isotype control to have internal controls. Phosphorylation is clearly decreased in a patient with STAT3 deficiency compared to a healthy control [45, 46].

See STAT phosphorylation staining protocol in Section 4.2.

4.2.3 STAT5 phosphorylation for the diagnosis of severe combined immunodeficiency (SCID)

X-linked severe combined immunodeficiency (X-SCID) is an inherited disorder of the immune system caused by inborn errors in the gene IL2RG, which encodes interleukin IL-2 receptor common γ chain. The γc is a transmembrane protein shared by different cytokine receptors IL-2, IL-4, IL-7, IL-9, IL-15, and IL-21. In the absence of an effective γc function, early lymphoid progenitor cells are not able to respond properly to these cytokine signals leading low or absence of T and NK cells and hypogammaglobulinemia despite a normal B cell count [47]. T−B+NK− phenotype is characteristic for X-SCID and JAK3 deficiencies. T−B+NK+ SCID include IL-7Rα, CD3δ, CD3ε, and CD3ζ deficiencies. X-SCID represents close to half of all SCID patients. Patients with X-SCID are prone to severe, recurrent, and persistent infections caused usually in their first 6 months of life. The disease is lethal within the first 2 years of life, unless they undergo hematopoietic stem cell transplantation (HSCT) or gene therapy [48, 49].

In addition to the more deleterious mutations, which cause the most severe phenotype of the disease, some other several patients with a milder form of combined immunodeficiency, often referred to as "leaky" or "hypomorphic" SCID. Most of these cases of "leaky" SCIDs are due to hypomorphic mutations. Due to their poorly defined clinical and immunological phenotype, patients with the atypical forms are diagnosed later in childhood or even in adulthood [50].

A study of expression of the common gamma chain (CD132), expressed broadly on T and B lymphocytes, NK cells, monocytes, and granulocytes, can be performed by flow cytometry [3]. However, as mentioned above, the presence of protein does not rule out a possible functional defect. It is therefore advisable to evaluate the signaling of the IL2-IL2RG-JAK3-STAT5 pathway, studying the phosphorylation of STAT5. For this purpose, whole blood is stimulated with IL-2 and STAT5 phosphorylation is evaluated in lymphocytes. It should be noted that a patient with X-SCID will not have T lymphocytes or NK cells, therefore the test should be performed on B lymphocytes and it is important to consider the inclusion of specific B cell lineage markers. STAT5 phosphorylation is clearly reduced in patients with a common gamma chain deficiency compared to a healthy control [39]. This assay can be performed under suspicion of mutations in the following genes: IL2RG, JAK3, STAT5 or IL-7Rα.

See STAT phosphorylation staining protocol in Section 4.2.

4.3 NK cell cytotoxicity and degranulation assay evaluation

Hemophagocytic lymphohistiocytosis (HLH) is a life-threatening condition characterized by persistent macrophage/lymphocyte activation and a multisystemic hyperinflammatory syndrome. The clinical diagnosis of HLH is challenging and requires a composite of several findings. According to the updated criteria of the Histiocyte Society [51], the diagnosis is currently established by fulfilling 5 of 8 clinical and laboratory criteria, or the finding of a genetic defect consistent with HLH. Defects associated with HLH are generally associated with abnormal NK cell function. The characterization of these patients requires genetic and functional studies, among the latter are the study of cytotoxic capacity and degranulation.

In *cytotoxicity assays*, we measure the functional capacity of natural killer cells. In this assay, PBMCs or purified NK cell are co-incubated at different ratios with a target HLA-null cell lines (usually K562) known to be sensitive to NK cell-mediated cytotoxicity [52]. After an incubation period, dead target cells are identified by nucleic acid staining, which specifically penetrates the dead cells and is measured by flow cytometry [53]. Chromium-51 (51Cr) release assays are also commonly used for accurate and precise quantification of cytotoxicity. The main disadvantages of this method are the use of radioactivity, which can be hazardous to health, and are impractical and costly due to the short half-life of 51Cr and to the need for radiological safety training and authorization requirements [54].

Cytotoxicity assay protocol:

1. Count and adjust patient PBMCs at 10^6 cells/mL concentration in a final volume of 5 mL of RPMI.

2. Count and adjust K562 cells at 2×10^5 cells/mL concentration in a final volume of 5 mL of RPMI.

3. Incubate both cell types separately at 37°C over night.

4. Count and adjust patient PBMCs at 2.5×10^6 cells/mL.

5. Count and adjust K562 cells at 0.2×10^6 cells/mL.

6. Make a serial dilution of the patient's PBMCs in FACS tubes, adding the same concentration of K562 cell line in each tube.

7. 5 μL of propidium iodide (PI) are added at each tube to identificate death cells through flow cytometry.

8. PBMcs mixed with K562 at different dilutions are incubated during 4 h at 37°C and in the dark.

9. Samples are ready to be acquired by flow cytometry.

Other lineage-specific function tests include evaluation of *NK cell degranulation* by evaluation of CD107a (LAMP1) surface expression, which is normally expressed in cytoplasmic granules, following stimulation with a target cell line (K562) or PMA plus ionomycin. This assay complements the classical NK cell cytotoxicity assay and is particularly useful in screening for the diagnosis of familial HLH. Specifically, the lack of surface expression of CD107a after incubation with K562 target cells is consistent with syntaxin-11 (STX11), uncoordinated mammals 13.4 (UNC13D), syntaxin-2 binding protein (STXBP2), or Rab27A protein defects [55].

NK cell degranulation assay protocol:

1. Count and adjust patient PBMCs at 10^6 cells/mL concentration.

2. Count and adjust K562 cells at 2×10^5 cells/mL concentration.

3. Dispense 100 μL of PBMCs at 10^6 cells/mL in a 96-well plate.

4. Leave three wells unstimulated by adding 100 μL of RPMI in each.

5. Add 100 μL of IL-2 to other three wells.

6. Incubate at 37°C over night.

7. Resuspend cell button and transfer it to FACS tubes.

8. Add 100 μL of RPMI into a tube unstimulated and into a IL-2 tube.

9. Add 100 μL of PHA into a tube unstimulated and into a IL-2 tube.

10. Add 100 μL of K26 (10^6 cells/mL) into a tube unstimulated and into a IL-2 tube.

11. Add 5 μL of anti-CD107a into all FACS tubes.

12. Incubate all FACS tubes during 3 h at 37°C.

13. Add the appropriate mix of surface antibodies of interest.

14. Incubate for 20 min at RT and in the dark.

15. Wash with 2 mL of PBS.

16. Discard the PBS after 5 min of centrifugation at 1500 rpm.

17. Add 100 μL of PBS and the sample is ready to be acquired by flow cytometry.

4.4 Detection of intracellular IFN-γ expression in activated T lymphocytes

The use of intracellular cytokine staining technique has become extremely useful in recent years to elucidate defects in intracellular signaling as causes of different IDPs. One of the applications of this technique is the characterization of the different subpopulations of T helper (Th) lymphocytes. The strategy of characterization of these subpopulations is to identify T helper 1 lymphocytes (Th1) as cells producing INFγ, T helper 2 lymphocytes (Th2) as those producing IL-4 and IL-13 and T helper 17 lymphocytes (Th17) as those producing IL-17a and IL-17f [1].

In patients with STAT3 or DOCK8 deficiency, a decrease in IL-17 producing T cells (Th17) has been demonstrated. Th17 cells play a role in autoimmunity and defense of extracellular pathogens (fungi, bacteria, and parasites). Therefore, the absence of IL-17 production after stimulation of PBMCs in these patients allows identifying a functional defect. However, mutational genetic analysis is necessary for a definitive diagnosis [56].

Moreover, this technique can be used to determine the production of other cytokines such as INFγ in front MSMD suspicion [57, 58] or the production of TNF-α in IRAK4 suspicion in defects of the innate immunity [3].

In summary, the intracellular cytokine staining technique consists of the stimulation of heparinized whole blood with PMA and ionomycin or other stimuli. After the appropriate incubation time for each cytokine, intracellular and superficial staining is performed. There are three aspects to consider: (A) the results will be compared with the cytokine production of a healthy control; (B) the incubation time with the stimulus can be modified depending on the cytokine we want to study. In case we want to study more than one, we will have to find an ideal incubation time for each of them, that is, leave time for all of them to be synthesized, without degrading any of them; (C) it is a key to know the cell subpopulation producing the cytokine of interest, as we can add surface markers that will facilitate analysis and interpretation (**Figure 9**).

Intracellular cytokine staining protocol:

1. Stimulate 500 μL of heparin whole blood with PMA and ionomycin.

2. Add 1 μL of GolgiPlug.

3. Incubate for 18 h at 37°C.

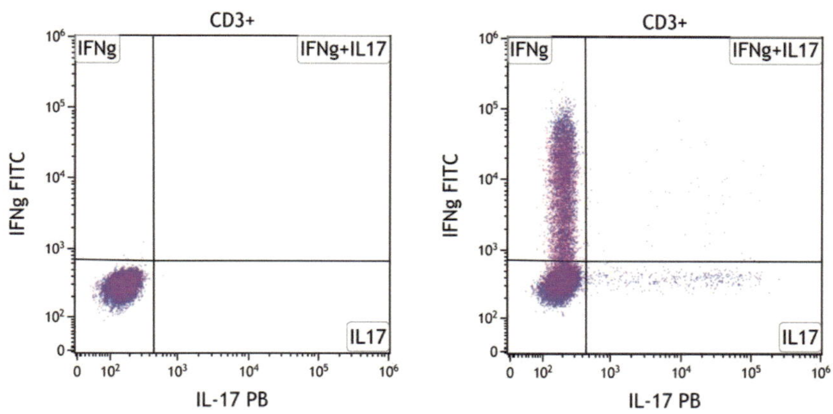

Figure 9.
Intracellular production of IFNg and IL2 cytokines. In the dotplots, cytokine production is observed after 18 h stimulation in a healthy donor (left) and a patient with an immune deficit (right).

4. Aliquot 200 μL of sample into different FACS tubes.

5. Wash with 2 mL of PBS.

6. Discard the PBS after 5 min of centrifugation at 1500 rpm (324 g).

7. Fix with the appropriate volume of fixative reagent.

8. Incubate for 15 min at RT and in the dark.

9. Add the appropriate volume of permeabilizing reagent.

10. Add the extracellular antibodies mix.

11. Add the intracellular antibodies mix.

12. Incubate for 20 min at RT and in the dark.

13. Add the appropriate volume of wash buffer.

14. Discard the wash buffer after 5 min of centrifugation at 1500 rpm.

15. Add 100 μL of PBS and the sample is ready to be acquired by flow cytometry.

(This protocol may change slightly depending on the commercial kit used).

5. Conclusions

Flow cytometry is an accessible technique in clinical laboratories; the implementation of this technology has progressed together with the significant improvements in instrumentation and the availability of a wide variety of monoclonal reagents. While the main clinical indications for immunophenotyping typically have been the quantification of CD4 T lymphocytes in HIV infection, the study of the extended immunophenotype and functional studies by flow cytometry offers nowadays advanced diagnostic approaches. Previously used techniques for the study of cell function (^{3}H-thymidine incorporation, ^{51}Cr-release assay, etc.) using radioactivity have been gradually replaced by flow cytometry. Flow cytometry can provide rapid and accurate identification of expanded lymphocyte subpopulations, which play an important role in the evaluation and understanding of complex disorders such as primary immunodeficiencies and other immunomediated diseases.

The application of flow cytometry techniques in the evaluation of PID assesses lymphocyte proliferation, intracellular changes associated with activation, cytokine production, and biological effects associated with immune defects and functional immune abnormalities.

Acknowledgements

We are grateful for the support of Immunology Division and Diagnostic Immunology Research Group from Hospital Universitario Vall d'Hebron and Vall d'Hebron Institut de Recerca (VHIR).

Conflict of interest

The authors declare no conflict of interest.

Author details

Mónica Martínez-Gallo[1,2,3]* and Marina García-Prat[3,4]

1 Immunology Division, Hospital Universitari Vall d'Hebron (HUVH), Vall d'Hebron Research Institute (VHIR), Barcelona, Catalonia, Spain

2 Department of Cell Biology, Physiology and Immunology, Universitat Autònoma de Barcelona (UAB), Barcelona, Catalonia, Spain

3 Jeffrey Modell Foundation Excellence Center, Barcelona, Spain

4 Pediatric Infectious Diseases and Immunodeficiencies Unit, Hospital Universitari Vall d'Hebron (HUVH), Vall d'Hebron Research Institute (VHIR), Universitat Autònoma de Barcelona (UAB), Barcelona, Catalonia, Spain

*Address all correspondence to: mmartinez@vhebron.net

References

[1] Abbas AK, Murphy KM, Sher A. Functional diversity of helper T lymphocytes. Nature. 1996;**383**(6603): 787-793

[2] Olbrich P, de Felipe B, Delgado-Pecellin C, Rodero R, Rojas P, Aguayo J, et al. A first pilot study on the neonatal screening of primary immunodeficiencies in Spain: TRECS and KRECS identify severe T- and B-cell lymphopenia. Anales de Pediatría. 2014;**81**(5):310-317

[3] Kanegane H, Hoshino A, Okano T, Yasumi T, Wada T, Takada H, et al. Flow cytometry-based diagnosis of primary immunodeficiency diseases. Allergology International. 2018;**67**(1):43-54

[4] Hsieh EWY, Hernandez JD. Novel tools for primary immunodeficiency diagnosis: Making a case for deep profiling. Current Opinion in Allergy and Clinical Immunology. 2016;**16**(6):549-556

[5] Carr EJ, Dooley J, Garcia-Perez JE, Lagou V, Lee JC, Wouters C, et al. The cellular composition of the human immune system is shaped by age and cohabitation. Nature Immunology. 2016;**17**(4):461-468

[6] Garcia-Prat M, Álvarez-Sierra D, Aguiló-Cucurull A, Salgado-Perandrés S, Briongos-Sebastian S, Franco-Jarava C, et al. Extended immunophenotyping reference values in a healthy pediatric population. Cytometry Part B: Clinical Cytometry. 2018;**96**(3):223-233

[7] Fleisher TA, Madkaikar M, Rosenzweig SD. Application of flow cytometry in the evaluation of primary immunodeficiencies. Indian Journal of Pediatrics. 2016;**83**(5):444-449

[8] Corneo B, Moshous D, Güngör T, Wulffraat N, Philippet P, Le Deist FL, et al. Identical mutations in RAG1 or RAG2 genes leading to defective V(D)J recombinase activity can cause either T-B-severe combined immune deficiency or Omenn syndrome. Blood. 2001;**97**(9):2772-2776

[9] Fink PJ. The biology of recent thymic emigrants. Annual Review of Immunology. 2013;**31**(1):31-50

[10] Sallusto F. Heterogeneity of human CD4(+) T cells against microbes. Annual Review of Immunology. 2016;**34**(1):317-334

[11] van den Broek T, Borghans JAM, van Wijk F. The full spectrum of human naive T cells. Nature Reviews. Immunology. 2018;**18**(6):363-373

[12] Farber DL, Yudanin NA, Restifo NP. Human memory T cells: Generation, compartmentalization and homeostasis. Nature Reviews. Immunology. 2014;**14**(1):24-35

[13] Azizi G, Rezaei N, Kiaee F, Tavakolinia N, Yazdani R, Mirshafiey A, et al. T-cell abnormalities in common variable immunodeficiency. Journal of Investigational Allergology & Clinical Immunology. 2016;**26**(4):233-243

[14] Dieli-Crimi R, Martínez-Gallo M, Franco-Jarava C, Antolin M, Blasco L, Paramonov I, et al. Th1-skewed profile and excessive production of proinflammatory cytokines in a NFKB1-deficient patient with CVID and severe gastrointestinal manifestations. Clinical Immunology. 2018;**195**:49-58

[15] Bacchetta R, Barzaghi F, Roncarolo M-G. From IPEX syndrome to FOXP3 mutation: A lesson on immune dysregulation. Annals of the New York Academy of Sciences. 2018;**1417**(1):5-22

[16] Warnatz K, Schlesier M. Flowcytometric phenotyping of common

variable immunodeficiency. Cytometry Part B: Clinical Cytometry. 2008;**74B**(5):261-271

[17] Wehr C, Kivioja T, Schmitt C, Ferry B, Witte T, Eren E, et al. The EUROclass trial: Defining subgroups in common variable immunodeficiency. Blood. 2008;**111**(1):77-85

[18] Avery DT, Kane A, Nguyen T, Lau A, Nguyen A, Lenthall H, et al. Germline-activating mutations in PIK3CD compromise B cell development and function. The Journal of Experimental Medicine. 2018;**215**(8):2073-2095

[19] Orange JS. Natural killer cell deficiency. The Journal of Allergy and Clinical Immunology. 2013;**132**(3):515-525

[20] Orange JS. Unraveling human natural killer cell deficiency. Journal of Clinical Investigation. 2012;**122**(3):798-801

[21] Nichols BA, Bainton DF, Farquhar MG. Differentiation of monocytes. Origin, nature, and fate of their azurophil granules. The Journal of Cell Biology. 1971;**50**(2):498-515

[22] Ziegler-Heitbrock L. The CD14+ CD16+ blood monocytes: Their role in infection and inflammation. Journal of Leukocyte Biology. 2007;**81**(3):584-592

[23] Banchereau J, Steinman RM. Dendritic cells and the control of immunity. Nature. 1998;**392**(6673):245-252

[24] Spinner MA, Sanchez LA, Hsu AP, Shaw PA, Zerbe CS, Calvo KR, et al. GATA2 deficiency: A protean disorder of hematopoiesis, lymphatics, and immunity. Blood. 2014;**123**(6):809-821

[25] Marcondes NA, Fernandes FB, Spindler BM, Faulhaber GAM. Flow cytometry assessment of intracellular BTK expression. Cytometry. Part B, Clinical Cytometry. 2018;**94**(5):568

[26] Gifford CE, Weingartner E, Villanueva J, Johnson J, Zhang K, Filipovich AH, et al. Clinical flow cytometric screening of SAP and XIAP expression accurately identifies patients with SH2D1A and XIAP/BIRC4 mutations. Cytometry. Part B, Clinical Cytometry. 2014;**86**(4):263-271

[27] Kogawa K, Lee SM, Villanueva J, Marmer D, Sumegi J, Filipovich AH. Perforin expression in cytotoxic lymphocytes from patients with hemophagocytic lymphohistiocytosis and their family members. Blood. 2002;**99**(1):61-66

[28] Demaret J, Saison J, Venet F, Malcus C, Poitevin-Later F, Lepape A, et al. Assessment of a novel flow cytometry technique of one-step intracellular staining: Example of FOXP3 in clinical samples. Cytometry. Part B, Clinical Cytometry. 2013;**84**(3):187-193

[29] Sava F, Toldi G, Treszl A, Hajdú J, Harmath Á, Tulassay T, et al. Expression of lymphocyte activation markers of preterm neonates is associated with perinatal complications. BMC Immunology. 2016;**17**(1):19

[30] Lim LC, Fiordalisi MN, Mantell JL, Schmitz JL, Folds JD. A whole-blood assay for qualitative and semiquantitative measurements of CD69 surface expression on CD4 and CD8 T lymphocytes using flow cytometry. Clinical and Diagnostic Laboratory Immunology. 1998;**5**(3):392-398

[31] O'Gorman MR. Measurement of CD40 ligand (CD154) expression on resting and in vitro-activated T cells. Current Protocols in Cytometry. 2001;Chapter 6:Unit 6.7

[32] Snyder JT, Shen J, Azmi H, Hou J, Fowler DH, Ragheb JA. Direct

inhibition of CD40L expression can contribute to the clinical efficacy of daclizumab independently of its effects on cell division and Th1/Th2 cytokine production. Blood. 2007;**109**(12):5399-5406

[33] Albert-Vega C, Tawfik DM, Trouillet-Assant S, Vachot L, Mallet F, Textoris J. Immune functional assays, from custom to standardized tests for precision medicine. Frontiers in Immunology. 2018;**16**(9):2367

[34] Oliveira JB, Fleisher TA. Laboratory evaluation of primary immunodeficiencies. The Journal of Allergy and Clinical Immunology. 2010;**125**(Suppl 2):S297-S305

[35] Lyons AB, Parish CR. Determination of lymphocyte division by flow cytometry. Journal of Immunological Methods. 1994;**171**(1):131-137

[36] Lašťovička J, Rataj M, Bartůňková J. Assessment of lymphocyte proliferation for diagnostic purpose: Comparison of CFSE staining, Ki-67 expression and 3H-thymidine incorporation. Human Immunology. 2016;**77**(12):1215-1222

[37] Casanova J-L, Holland SM, Notarangelo LD. Inborn errors of human JAKs and STATs. Immunity. 2012;**36**(4):515-528

[38] Bitar M, Boldt A, Binder S, Borte M, Kentouche K, Borte S, et al. Flow cytometric measurement of STAT1 and STAT3 phosphorylation in CD4 + and CD8 + T cells—Clinical applications in primary immunodeficiency diagnostics. The Journal of Allergy and Clinical Immunology. 2017;**140**(5): 1439-1441.e9

[39] Lorenzini T, Dotta L, Giacomelli M, Vairo D, Badolato R. STAT mutations as program switchers: Turning primary immunodeficiencies into autoimmune diseases. Journal of Leukocyte Biology. 2017;**101**(1):29-38

[40] Yanagihara N, Tachikawa E, Izumi F, Yasugawa S, Yamamoto H, Miyamoto E. Staurosporine: An effective inhibitor for Ca^{2+}/calmodulin-dependent protein kinase II. Journal of Neurochemistry. 1991;**56**(1):294-298

[41] Mizoguchi Y, Tsumura M, Okada S, Hirata O, Minegishi S, Imai K, et al. Simple diagnosis of STAT1 gain-of-function alleles in patients with chronic mucocutaneous candidiasis. Journal of Leukocyte Biology. 2014;**95**(4):667-676

[42] Depner M, Fuchs S, Raabe J, Frede N, Glocker C, Doffinger R, et al. The extended clinical phenotype of 26 patients with chronic mucocutaneous candidiasis due to gain-of-function mutations in STAT1. Journal of Clinical Immunology. 2016;**36**(1):73-84

[43] Chapgier A, Wynn RF, Jouanguy E, Filipe-Santos O, Zhang S, Feinberg J, et al. Human complete Stat-1 deficiency is associated with defective type I and II IFN responses in vitro but immunity to some low virulence viruses in vivo. Journal of Immunology. 2006;**176**(8):5078-5083

[44] Holland SM, DeLeo FR, Elloumi HZ, Hsu AP, Uzel G, Brodsky N, et al. STAT3 mutations in the hyper-IgE syndrome. The New England Journal of Medicine. 2007;**357**(16):1608-1619

[45] Meshaal SS, El Hawary RE, Eldash A, Grimbacher B, Camacho-Ordonez N, Abd Elaziz DS, et al. Diagnosis of DOCK8 deficiency using flow cytometry biomarkers: An Egyptian center experience. Clinical Immunology. 2018;**195**:36-44

[46] Renner ED, Rylaarsdam S, Anover-Sombke S, Rack AL, Reichenbach J, Carey JC, et al. Novel signal transducer and activator of transcription 3 (STAT3) mutations, reduced T(H)17 cell numbers, and variably defective

STAT3 phosphorylation in hyper-IgE syndrome. The Journal of Allergy and Clinical Immunology. 2008;**122**(1):181-187

[47] Leonard WJ, Noguchi M, Russell SM, McBride OW. The molecular basis of X-linked severe combined immunodeficiency: The role of the interleukin-2 receptor gamma chain as a common gamma chain, gamma c. Immunological Reviews. 1994;**138**:61-86

[48] Noguchi M, Yi H, Rosenblatt HM, Filipovich AH, Adelstein S, Modi WS, et al. Interleukin-2 receptor gamma chain mutation results in X-linked severe combined immunodeficiency in humans. Cell. 1993;**73**(1):147-157

[49] Kovanen PE, Leonard WJ. Cytokines and immunodeficiency diseases: Critical roles of the gamma(c)-dependent cytokines interleukins 2, 4, 7, 9, 15, and 21, and their signaling pathways. Immunological Reviews. 2004;**202**(1):67-83

[50] Mou W, He J, Chen X, Zhang H, Ren X, Wu X, et al. A novel deletion mutation in IL2RG gene results in X-linked severe combined immunodeficiency with an atypical phenotype. Immunogenetics. 2017;**69**(1):29-38

[51] Janka GE, Lehmberg K. Hemophagocytic syndromes—An update. Blood Reviews. 2014;**28**(4): 135-142

[52] Lisovsky I, Isitman G, Bruneau J, Bernard NF. Functional analysis of NK cell subsets activated by 721.221 and K562 HLA-null cells. Journal of Leukocyte Biology. 2015;**97**(4):761-767

[53] Kandarian F, Sunga GM, Arango-Saenz D, Rossetti M. A flow cytometry-based cytotoxicity assay for the assessment of human NK cell activity. Journal of Visualized Experiments. 2017;**9**(126):56191-56199

[54] Kim GG, Donnenberg VS, Donnenberg AD, Gooding W, Whiteside TL. A novel multiparametric flow cytometry-based cytotoxicity assay simultaneously immunophenotypes effector cells: Comparisons to a 4 h 51Cr-release assay. Journal of Immunological Methods. 2007;**325**(1-2):51-66

[55] Rubin TS, Zhang K, Gifford C, Lane A, Choo S, Bleesing JJ, et al. Perforin and CD107a testing is superior to NK cell function testing for screening patients for genetic HLH. Blood. 2017;**129**(22):2993-2999

[56] Sandquist I, Kolls J. Update on regulation and effector functions of Th17 cells. F1000Research. 2018;**7**:205

[57] Esteve-Solé A, Sologuren I, Martínez-Saavedra MT, Deyà-Martínez À, Oleaga-Quintas C, Martinez-Barricarte R, et al. Laboratory evaluation of the IFN-γ circuit for the molecular diagnosis of Mendelian susceptibility to mycobacterial disease. Critical Reviews in Clinical Laboratory Sciences. 2018;**55**(3):184-204

[58] Feinberg J, Fieschi C, Doffinger R, Feinberg M, Leclerc T, Boisson-Dupuis S, et al. Bacillus Calmette Guerin triggers the IL-12/IFN-gamma axis by an IRAK-4- and NEMO-dependent, non-cognate interaction between monocytes, NK, and T lymphocytes. European Journal of Immunology. 2004;**34**(11):3276-3284

Chapter 3

Flow Cytometry Assay for Quantitation of Therapeutical Anti-D IgG during Process Control in the Pharmaceutical Production

Sergio A. Oviedo

Abstract

Individuals who do not possess the D antigen in their red blood cells generate Anti-D antibodies against an antigenic challenge. Prophylaxis with Anti-D immunoglobulin prevents sensitization. The determination of adequate doses of Anti-D in plasma and pharmaceutical products is carried out by radio immuno assay (RIA) and enzyme immuno assay (EIA) or hemagglutination. An in house technique was developed for the quantitation of Anti-D antibodies, as an alternative test to the reference method. It was specific and with good recovery and did not present false positives or autoagglutination. The dose-response curve (mean fluorescence intensity (MFI) versus logarithm of concentration (log C)) was linear (correlation coefficient of 0.99). The method was validated following the criteria of the NIBSC (National Institute of Biological Standards and Control) and the European Pharmacopoeia. Flow cytometry allowed obtaining accurate, precise, sensitive and specific determinations at different concentrations in different biological matrices. The method can be used in highly diluted samples, has a strong fluorescence signal, is simple, fast, reliable and of relatively low cost. Flow cytometry is more efficient than hemagglutination and easier than RIA. With similar security and efficiency standards, it is cheaper than EIA an RIA. This method as a more suitable choice.

Keywords: cytometry assay, Anti-D IgG (Rho), process control, industrial process

1. Introduction

The Anti-D immunoglobulin (Anti-D IgG) produced from human plasma is used in the immunoprophylaxis of hemolytic disease of the newborn individuals who do not possess the red blood cells D antigen. Health authorities and manufacturers determine the concentration of antibody for each batch of medicine. Initially, the monitoring of Anti-D in the plasma of pregnant women and in finished products (Anti-D IgG) was performed by the anti-globulin test or Coombs test [1, 2] or by the continuous flow analysis of hemagglutination with a Technicon Autoanalyzer [3–8]. The automatic hemagglutination method has been considered as a reference by the European Pharmacopeia (EP); despite this, some groups of experts and regulatory bodies have replaced such methodology by radioimmunoassay (RIA)

[9–11] flow cytometry and a competitive enzyme immunoassay (EIA). These alternatives to the current reference method of European Pharmacopeia have been successfully investigated during international collaborative studies [11–17]. Competitive flow cytometry and EIA assays have been described previously [18–20].

We describe the use of flow cytometry in the control of manufacturing processes of Anti-D globulin, establishing the validation parameters and compliance with the quality standards required by the GMP and established by the producer at the different stages of the industrial process (hyperimmune plasma, Fraction II of the Cohn-Oncley process, and finished product) [13, 14, 21–23].

The rationale of the technique lies in the binding antigen antibodies, which are labeled with fluorescein detected by flow cytometry. The Anti-D immunoglobulin is quantified in comparison with the international reference preparation, calibrated in international units (IU), which allows to give a specification in IU/ml [20, 24]. In relation to flow cytometry, Expert Group No 6B of the NIBSC standardized a procedure to be applied in the evaluation of Anti-D in Anti-D IgG solutions [24]. In our laboratory, a flow cytometry technique was developed for the quantititation of Anti-D antibodies of the Gamma-Rho UNC, designed according to the procedures described in the literature [20, 24] and validated according to the criteria indicated by the NIBSC expert group [16, 24].

2. Materials and methods

2.1 Description of the flow cytometry test

A laser hits the red fluorescein-labeled cells that cross the cytometer in a continuous flow. The light emission produced by the interaction is detected by the forward scatter (FS) and converted into a voltage pulse (FS). A limit of 25,000 events is determined at a flow rate of 600 cells per second with a power of 1300 V. The concentration of Anti-D will be proportional to the size and intensity of fluorescence emitted by the antigen-antibody fluorescein complex of each red blood cell. The mean fluorescence intensity is plotted according to the logarithm of the Anti-D concentration.

2.2 Preparation of cell suspension and Anti-D IgG samples and sensitization and immunoglobulin binding to the cell surface

A Rho red cell suspension (Red Cell phenotypes R1R1/R1R2 and Rh-rr as control), were diluted to 0.2% in phosphate-buffered saline (PBS) pH 72 and plasma samples, Fraction II and finished product were diluted 1/5 to 1/100 with 1% PBS—pH 7.2.

2.3 Sensitization and immunoglobulin binding to the cell surface

0.2% red cell suspension was incubated with different test fractions (Anti-D standard IRP 68/419 OMS-NIBSC, human plasma, Fraction II of the Cohn process and gamma Anti-D globulin) in a 50/50 μl ratio. After stirring 30 min at 4°C, red blood cell suspension is washed with PBS-1% BSA and incubated with a 1/50 dilution of a goat anti-human IgG FITC (**Figure 1**). Red blood cells resuspended in 1 ml of PBS-1% BSA are measured with flow cytometry in Epics XL-MCL Cytometer-Coulter.

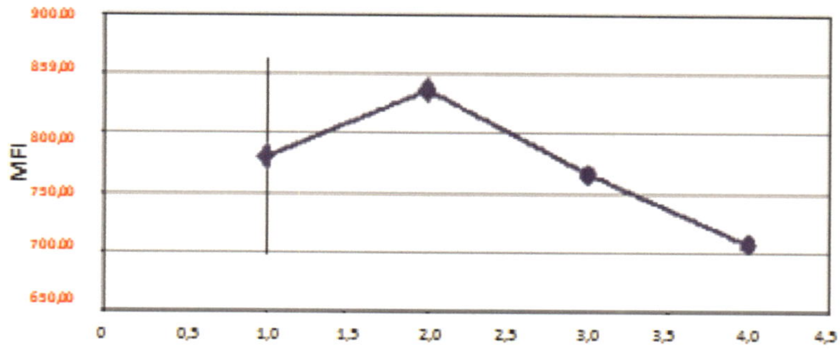

Figure 1.
Calibration curve of the conjugate dilutions.

2.4 Standard curve and statistical treatment

In a concentration range of 1500 ng/ml (7.5 IU/ml)–180 (0.9 IU/ml) ng/ml of international reference for human Anti-D immunoglobulin preparation (IRP 68/419 WHO), we establish the standard curve. No Anti-D sensitized cells were used, to set false positives and the minimum level of detection. Mean fluorescence intensity values were between 600 and 1900 and averaged around 800. The upper and lower limit values were determined using five standard curves. Nonparametric statistics was used and processed with Microsoft Excel 7.0 program and Method Validator 1.15 [25–28].

3. Results

3.1 Calibration of the xytometer

Epics XL-MCL Cytometer (Coulter, Corp., Luston, UK) was checked and calibrated using the standard immunocheck particles (Becton Dickinson, Oxford, UK). The red cell samples automatically passed through the cytometer according to their FS and side scatter (SS). After defining the working conditions and the protocol to be used with the cytometer, it was determined that the most suitable conjugate dilutions were 1/50 and 1/100 (**Figure 1**). It was determined to work with 25,000 events at a flow rate of 600 events/s (**Figure 2**). The intensity of the signal observed in the negative controls is related to non-specific negative unions. There were no other phenomena related to false positives (**Figure 3**).

The most suitable conjugate was determined by processing 26 tubes in duplicate and 1/100 dilutions of the conjugate (Sigma Anti Human IgG -Fc Conjugate-F9512 specific and Kallestad FITC Conjugate #30446). Serial dilutions of commercial gamma globulin (250 μg/ml) were used as control. **Figure 4** shows that the Sigma conjugate presents a greater fluorescence signal. After 10 assays using different fluorescence particles, the fluorescence intensity (MESF) and the fluorescence signal emitted by the cytometer presented good correlation. **Figure 5** shows the linear relationship between the emitted MFI and the MESF. Peak's SD values in a range of 1.4–7.6 represent points with higher and lower intensity.

The histograms of the fluorescence parameters that were plotted according to the anti-Rho IgG concentrations can be seen in **Figure 6**. **Figure 7a** shows the working area determined with the forward and lateral dispersion of a homogenous population of non-sensitized cells. We observed no significant agglutination of auto aggregated red cells for anti-Rho IgG concentrations less than or equal to 960 ng/ml (4.8 IU/ml). We worked in intervals of ≤960 ng/ml to avoid increments of FS caused by auto aggregation (**Figure 7b**).

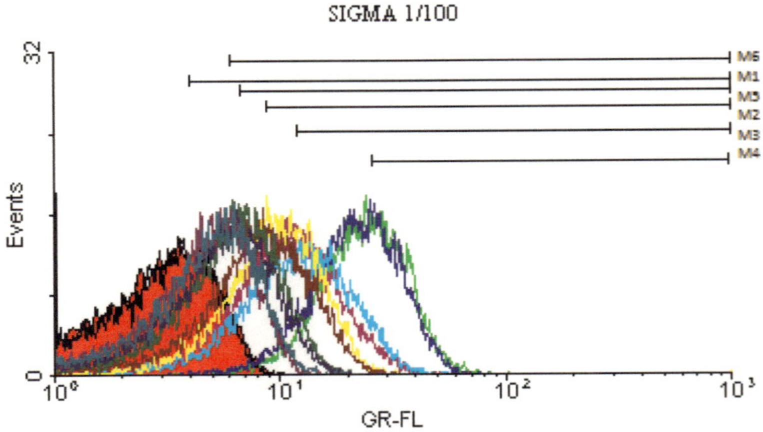

Figure 2.
Work and flow rate events.

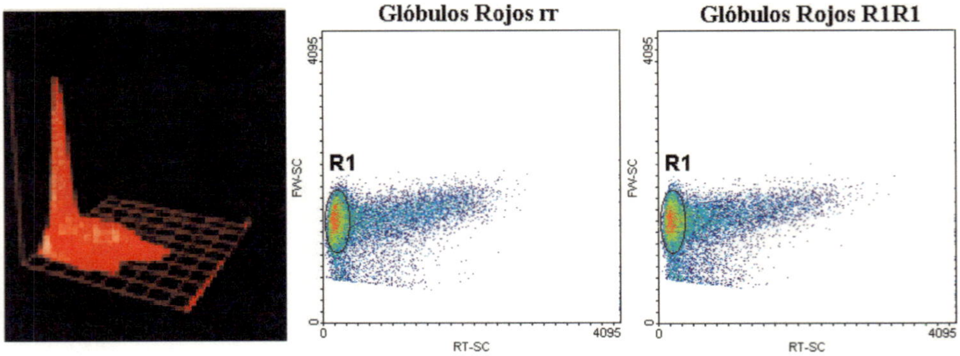

Figure 3.
Forward (FS) and side (SS) scatter graphs of a homogeneous population of non-sensitized cells.

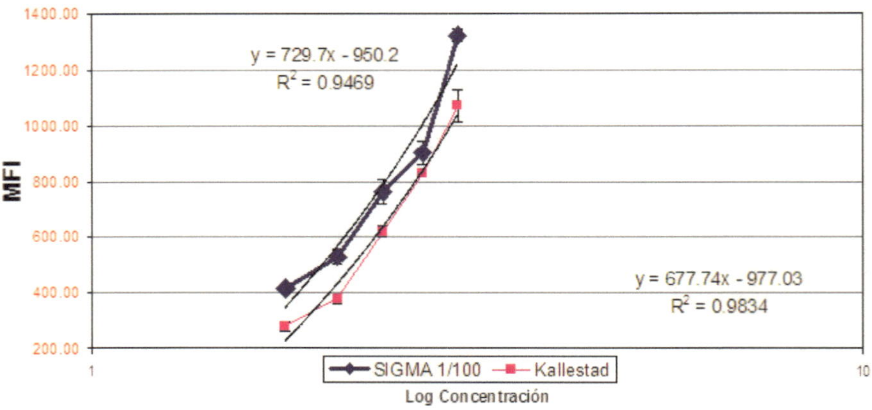

Figure 4.
Comparative curve of the conjugates.

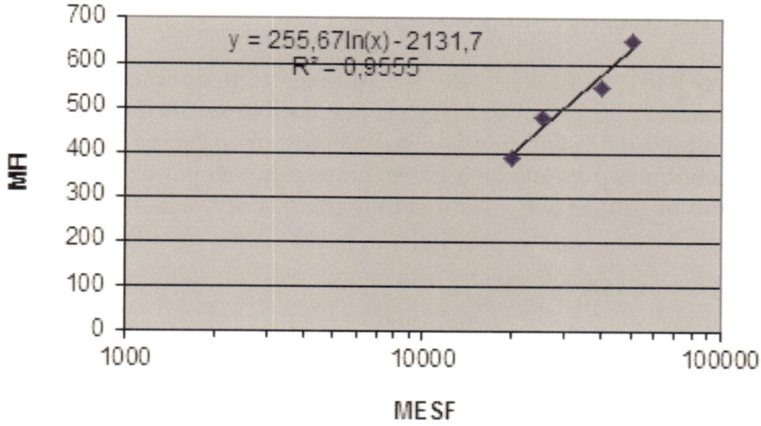

Figure 5.
Calibration curve of the flow cytometer, correlation between MFI and MESF.

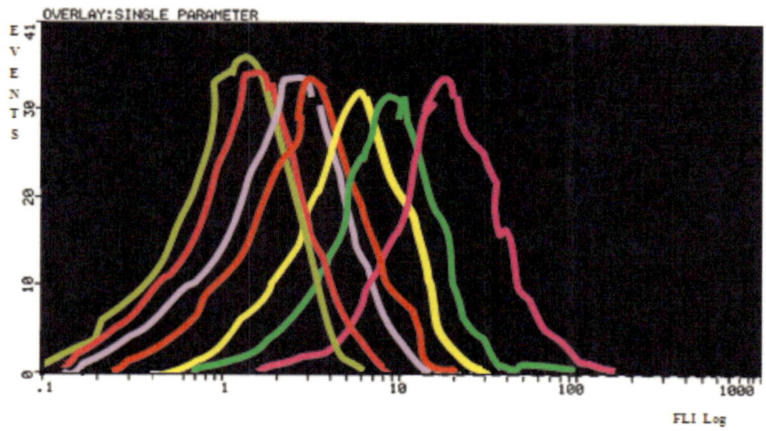

Figure 6.
FL1 histogram of fluorescence stockings obtained from seven dilutions of the WHO 68/419 standard using GR Rho + R1R1, ------- 960 ng/ml AD, ------- 480 ng/ml AD,--- 240 ng/ml AD, ------- 120 ng/ml AD, ----- 60 ng/ml AD, and ------- 15 ng/ml AD.

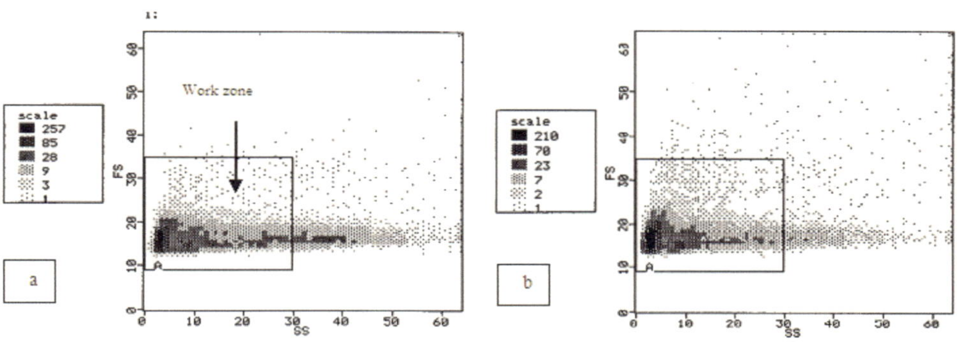

Figure 7.
(a) Working region selected for Rho + R1R1 red blood cells according to the FS and SS of washed cells not sensitized in a concentration of 0.2% in PBS. (b) Increase of FS of GR R1R1 sensitized with anti-Rho IgG standard 68/419 960 ng/ml.

3.2 Effect of the red blood cell phenotype

Phenotypes R1R1, R1R2, R1r′, and R1r were tested to optimize the assay. The values of estimated concentrations (means-standard deviations-CV%) suggest that the tests are reproducible for each sample and each phenotype (**Table 1**). The R1R1 and R1R2 phenotypes presented a greater fluorescence signal, and, therefore, a greater amount of Anti-D was adsorbed than the rest of the phenotypes (**Figure 8**).

3.3 Standard curve and cutoff

To establish the standard curve, dilutions of WHO reference preparation were used in a concentration range of 30–480 ng/ml. The corresponding values of MFI were between 1.4 and 4.5. To establish the minimum level of detection, non-sensitized cells were run, with MFI values averaging 1.2 (**Figure 9**).

A linear and proportional to the log of the Anti-D concentration fluorescence response was obtained in the measured range. The linear regression of 270 points was r = 0.963. The linear measurement range was set between 120 and 240 ng/ml and was the most acceptable for our purpose.

Seven similar curves were run with each other (**Figure 10a**), and the average curve and two SD values were calculated at each point, and the upper and lower

Phenotype		**R1R1**			**R1R2**	
ng STD	480	120	60	480	120	60
ng Rec.	480.45	119.70	60.30	478.75	118.00	60.10
	475.80	118.98	60.00	477.30	119.00	56.90
	481.06	120.00	59.45	476.89	118.5	59.89
	480.03	121.00	60.10	477.00	116.78	60.05
Media	479.33	119.92	59.96	477.48	118.07	59.235
SD	2.39	0.84	0.36	0.86	0.95	1.56
VC%	0.5	0.7	0.6	0.2	0.8	2.6
SD + 2	484.12	121.59	60.69	479.21	119.97	62.35
SD − 2	474.54	118.24	59.23	475.76	116.17	56.12
Phenotype		**R1r′**			**R1r**	
ng STD	480	120	60	480	120	60
ng Rec.	476.35	117.45	60.10	465.98	120.21	60.10
	469.40	117.67	56.90	475.35	114.78	56.90
	474.45	118.25	57.98	474.95	118.25	57.98
	475.60	117.33	58.87	477.60	117.33	56.89
Media	473.95	117.675	58.46	473.47	117.64	57.96
SD	3.13	0.40	1.35	5.12	2254	1.51
VC%	0.7	0.3	2.3	1.08	1.9	2.6
SD + 2	480.21	118.49	61.17	483.72	122.15	60.98
SD − 2	467.68	116.85	55.74	463.21	113.13	54.94

Table 1.
Red blood cell phenotype effect.

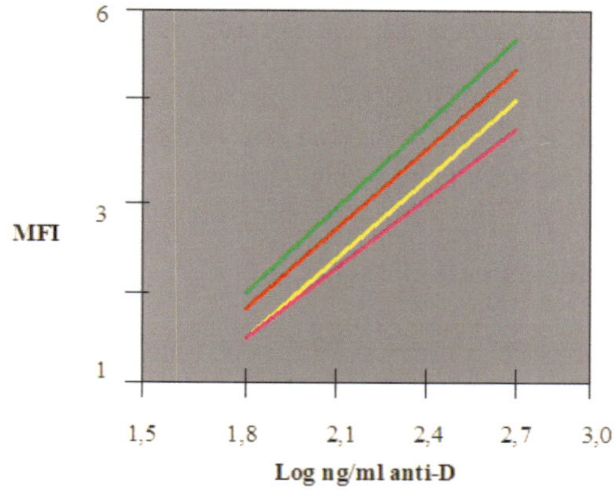

Figure 8.
Standard curves using red blood cells of phenotype: R1R1 (-----), R1R2 (------), R1r´ (------), and R1r (------).

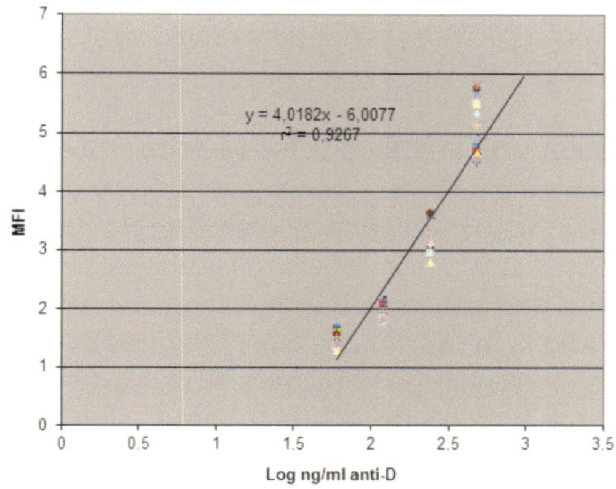

Figure 9.
Construction of the standard curve. A total of 18 curves was run and its linear regression was r = 0.963.

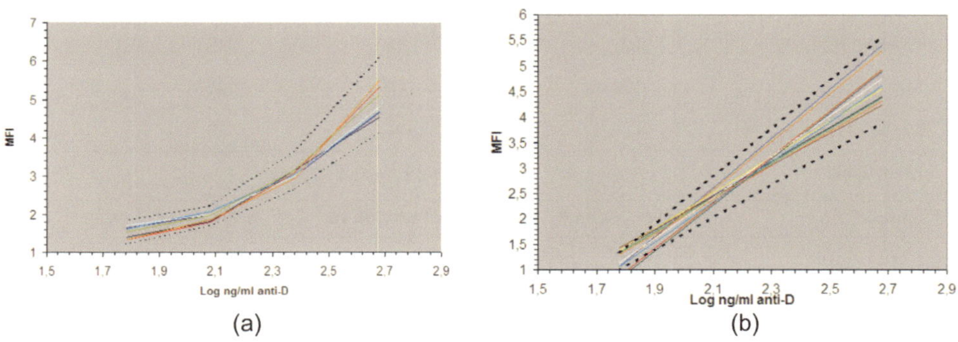

Figure 10.
(a) Determination of the lower and upper limits (------) and the values of ±2DS for the mean standard curve. The limit lines were used for the approval of the standard curves. (b) From 19 standard curves, 15 were found completely within the lower and upper limit lines.

limits of each reading were calculated and plotted. To determine the appropriate limits for our purpose, 19 standard curves were tested. As shown in **Figure 10**, at high concentrations of Anti-D, there are no significant deviations, and in the diluted samples, three showed deviations from the upper limit (0.24 μg/ml) and one below the lower limit (120 ng/ml). The limits of 120–240 ng/ml were taken as validation criteria of the standard curve.

The cutoff value set at 10 ng/ml (0.05 IU/ml) corresponds to a 1:1024 Anti-D standard (NIBS) dilution. After 135 plasma assays (55 Anti-D positive—80 negative), it was determined.

3.4 Reproducibility and repeatability

Tests in different matrices repeated 10 times simultaneously (intra-assay) during different days (inter-assay) evaluated the reproducibility and repeatability of flow cytometry. The sources of variation found were linked to the preparation of samples and variations in the cytometer (**Table 2**). The test showed good repeatability between tests, indicated by the coefficient of variation in representative samples (CV 4.95%, high concentration; 3.36%, intermediate concentration; and 4.78%, low concentration) (**Table 3**). These results coincide with Thorpe and Schäffner in the collaborative study of the European Union [13].

3.5 Recovery of Anti-D

The estimated concentration of Anti-D for each measured point in three commercial products presented a %CV in the range of 1–7.5% (**Table 4**). Two-thirds of the powers estimated by the manufacturers were within 95% of the acceptance limit of flow cytometry.

Only 1 of the 30 tests performed on 3 known concentration of commercial products showed deviations in the cut of the determined concentrations (87.3–111.9). The assay presented good recoveries of concentration in the samples tested (**Table 4**). The concentrations were estimated using the Wilcoxon test.

Intra-assay	**(a) Reference standard 68/419 WHO in ng/ml**			
Concentration	480	120	30	
n	10	10	10	
Average MFI	4.79	1.98	1.25	
SD	0.24	0.07	0.06	
VC %	4.95	3.36	4.78	
Intra-assay	**(b) Sample control µg/ml**			
Sample	**Pool of plasma***	**Cohn Fraction II ***	**γ1***	**γ2***
Concentration	12.5	30.2	197.2	179.5
n	10	10	10	10
Average MFI	2.08	11.26	3.12	3.49
SD	0.06	0.05	0.16	0.17
VC%	3.06	4.32	5.16	4.93
% Recovery	113.6	91.56	98.33	109.46

**Commercial gamma globulins: concentration 1 = 204 µg/ml, 2 = 164 µg/ml; pool = 11 µg/ml, Cohn's Fraction II = 33 µg/ml.*

Table 2.
Accuracy and reproducibility of the standard (a) and control samples (b).

Inter-assay: (a) reference standard 68/419 WHO in ng/ml											
ng/ml	**MFI**							**SD**	**VC%**	**+2SD**	**−2SD**
	M 1	**M 2**	**M 3**	**M 4**	**M 5**	**M 6**	**_ X**				
480	5.95	5.12	5.06	5.44	4.76	5.22	5.26	0.40	7.69	8.38	4.45
240	3.11	3.17	3.21	2.95	3.01	3.34	3.13	0.14	4.49	5.57	2.85
120	1.8	1.97	1.88	1.84	2.11	2.01	1.93	0.12	6.01	3.97	1.70
60	1.31	1.53	1.39	1.31	1.63	1.59	1.46	0.14	9.71	1.46	1.18
30	1.11	1.23	1.27	1.22	1.3	1.2	1.22	0.06	5.36	1.22	1.09

Inter-assay: (b) Control samples									
Product	**Gamma H***			**Gamma P***			**Gamma B***		
Dilution	1/250	1/500	1/750	1/350	1/700	1/1000	1/250	1/500	1/1000
MFI	1.21	1.16	1.05	1.37	1.24	1.06	1.21	1.15	1.02
	1.18	1.16	1.04	1.39	1.24	1.06	1.23	1.15	1.03
	1.19	1.17	1.04	1.39	1.25	1.06	1.19	1.13	0.98
	1.22	1.19	1.04	1.38	1.23	1.07	1.22	1.14	1.00
	1.20	1.14	1.06	1.3	1.15	1.05	1.21	1.16	1.01
	1.18	1.15	1.05	1.27	1.22	1.04	1.23	1.13	1.02
Average MFI	1.19	1.161	1.05	1.35	1.22	1.06	1.215	1.14	1.01
SD	0.016	0,017	0,008	0.05	0.04	0.01	0.015	0.012	0.018
VC%	1.36	1.48	0.78	3.83	2.99	0.98	1.25	1.06	1.77
µg/ml	149.50	150.25	142.64	236.25	220.86	192.00	151.87	147.41	183.64
% Recovery	99.66	100.16	95.09	107.39	100.39	87.27	92.61	89,89	111.97
+2SD	1.23	1.20	1.06	1.45	1.29	1.08	1.24	1.17	1.04
−2SD	1.16	1.13	1.03	1.25	1.15	1.04	1.18	1.12	0.97

**Declared concentration for H: 150 µg/vial, P: 220 µg/vial, B: 164 µg/vial.*

Table 3.
Accuracy and repeatability of standard (a) control samples (b).

Product	n	Average µg/ml	Limit 95%	VC%	Declared concentration µg/ml	% Recovery
H	6	149.50	145–154	2.6	150	99.7
	6	150.25	146–155	3.3		100.2
	6	142.64	140–144	1.3		95.1
P	6	236.25	218–254	7.2	220	107.4
	6	220.86	208–233	5.7		100.4
	6	192.00	189–196	2.1		87.3
B	6	151.87	147–155	1.9	164	92.6
	6	147.1	145–151	2.7		89.9
	6	183.64	176–189	2.7		111.9

Table 4.
Comparison between anti-D concentration declared by the producer and concentration obtained by flow cytometry.

4. Discussion

The availability of a quantitation method for human plasma, Cohn fractions, and Anti-D gammaglobulin has a relevant importance for the process control of the Anti-D IgG production. Pharmaceutical companies are employing sensitive EIA, RIA, and flow cytometry techniques [10, 11, 15, 20, 24] for the quantitation of Anti-D level in finished products, the use in Anti-D gammaglobulin manufacturing process control have no be reported.

A rapid and low-cost flow cytometry test applied in the control of the industrial process of large-scale production obtaining Anti-D gammaglobulin has been described. This responded to the need of an alternative method to EIA and RIA in the quantitation of Anti-D in different process samples (human plasma, Chon fractions, gammaglobulin) [29].

According to this, the threshold of the assay was adjusted to select the level of antibody content, which will be accepted, to produce an immunoglobulin with a potency of 250 μg/vial (1250 IU/vial), according to the regulatory requirements [13, 14, 28]. To validate the assay, chemical and physical-chemical factors that affect the binding of antigen-antibody influence in the red blood cells were determined. The spatial conditions of cell packing did not affect the assay. The adsorption of antibodies to the cell surface was 1000 times greater when the incubation temperature was 37°C without this affecting the dissociation rate [9]. The different fractions behave similarly in the assay conditions, and there was no interference by autoagglutination at the concentration of red blood cells used (**Figure** 7). The analytical quality assurance of the trial showed that the method presents good reproducibility, repeatability (1–7.5 VC%), and correlation with the standard curve (r: 0.9267). The assay is specific and recovery is 95% of the known Anti-D concentration values. For validation our strategy was based on an international guide accepted by manufacturers and control laboratories [13, 30]. The validation method meets the requirements according to the good manufacturing practices in the pharmaceutical industry. A validation methodology was followed to meet the special requirements of standardization of the good manufacturing practices in the pharmaceutical industry [30, 31].

Biological variability, red blood cells, and matrices were considered in a validation protocol according to international guidelines [14, 30].

5. Conclusions

This flow cytometry method is a sensitive and specific method that allows reproducible results. The concentration values are comparable with those estimated with other methods such as RIA and EIA used by commercial manufacturers of gamma globulins, and the method presents accuracy and precision. The test is completed in 3 h and is easy to perform, allowing quantitative assessment of Anti-D antibodies from plasma, fractions of the Cohn process, and finished products. The flow cytometry method presented shorter processing times (3 h) than the EIA or RIA methods (5 h).

The potential of the flow cytometry method described here represents an alternative to quantify Anti-D in different matrices and meets the criteria of good laboratory practices (simple, fast, and reliable as well as being sensitive and accurate). In the control of industrial processes, this method has shown reproducibility and reliability for this purpose.

The cost of materials is slightly lower for flow cytometry ($ 15/test) than in EIA or RIA ($ 25/test).

Highly significant correlations were observed between the flow cytometry test Anti-D values and EIA values determined in different matrices (Serum, Fraction II, Semi-elaborated, gamma globulin).

For the case of process samples (Fraction II, semi-processed), strong correlations were observed between the flow cytometry test and the EIA values. A potential advantage of the flow cytometry assay could be the higher sensitivity presented than the EIA assay.

It is concluded that the flow cytometry method has advantages over the EIA and RIA method as a substitute for the present standard method.

Author details

Sergio A. Oviedo[1,2]

1 Blood Products Laboratory National University of Cordoba, Córdoba, Argentina

2 Group Quality and Regulatory Affairs, School of Chemistry, Catholic University of Cordoba, Argentina

*Address all correspondence to: sergio.oviedo@unc.edu.ar

References

[1] Coombs RRA, Mourant AE, Race RR. Detection of weak and 'incomplete' Rh agglutinins: A new test. The Lancet. 1945;**246**(6358):15-16

[2] Coombs RRA, Mourant AE, Race RR. A new test for the detection of weak and 'incomplete' Rh agglutinins. British Journal of Experimental Pathology. 1945;**26**:255-266

[3] Rosenfield RE, Harber GV. Detection and measurement of homologous human hemagglutinins. In: Technicon Symposia, Automation in Analytical Chemistry. 1966. pp. 503-506

[4] Moore BPL, Hughes-Jones NC. Automated assay of anti-D concentration in plasmapheresis donors. In: Proceedings Technicon International Congress; Chicago. 1970

[5] Perrault R, Hogman C. Automated red cell antibody analysis: A parallel study. I. Detection and quantitation. Vox Sanguinis. 1971;**20**:340-355

[6] Esposito VM, Paul EB, Seligmann EB. In vitro potency results from a collaborative study of Rh,(D) immune globulin reference preparation. Vox Sanguinis. 1971;**21**:193-199

[7] Gunson HH, Philips PK, Statton F. Observations on the reproducibility of the bromelised test cell anti-D assay, using the AutoAnalyser. Vox Sanguinis. 1976;**26**(334–343):1976

[8] Paul EB. Standardization of US reference Rho (D) immune globulin by quantitative automated hemagglutination. Journal of Biological Standardization. 1986;**14**:121-125

[9] Hughes-Jones NC, Gardner B, Telford R. Comparison of various methods of dissociating anti-D, using 1311-labelled antibody. Vox Sanguinis. 1963;**8**:531-536

[10] Hughes-Jones N, Dawes B, Gorick B, Melamed M, Pepper R, Thompson K. Radio-immunoassay of the functional activity of anti-D (Rh) preparations using a human monoclonal ^{125}I-Labelled anti-D. Vox Sanguinis. 1987;**53**:175-180. DOI: 10.1111/j.1423-0410.1987.tb04945.x

[11] Bangham DR, Kirkwood TBL, Wybrow G, Hughes-Jones NC, Gunson HH. International collaborative study of assay of anti-D (anti-Rho) immunoglobulin. British Journal of Haematology. 1978;**38**:407-423

[12] Gunson HH, Bowell PJ, Kirkwood TBL. Collaborative study to recalibrate the international reference preparation of anti-D immunoglobulin. Journal of Clinical Pathology. 1980;**33**:249-253

[13] Anonymous: Assay of human anti-D immunoglobulin 2-7-13. In: European Pharmacopoeia. 4th ed. Strasbourg: European Commission; 2002

[14] Anonymous: 2.7.13 Assay of Human anti-D Immunoglobulin: European Pharmacopoeia. 4th ed. Strasboung: Directorate for the Quality of Medicines of the Council of Europe; 2003 [addendum 4·6:3890-3892]

[15] Austin EB, McIntosh Y, Hodson C, Lee D. anti-D quantitation by flow cytometry: An alternative to the autoanalyzer? Transfusion Medicine. 1995;**5**:203-208

[16] Thorpe SJ, Turner CE, Heath AC, Sands D. A competitive enzyme-linked immunoassay using erythrocytes fixed to microtitre plates for anti-D quantitation in immunoglobulin products. Vox Sanguinis. 2000;**79**:100-107

[17] Thorpe SJ, Sands D, Rautmann G, Schäffner G. International collaborative

study to evaluate methods for quantification of anti-D in immunoglobulin preparations. Vox Sanguinis. 2002;**83**:42-50

[18] Thorpe SJ, Fox B, Sands D. A stable lyophilized reagent for use in a potential reference assay for quantitation of anti-D in immunoglobulin products. Biologicals. 2002;**30**:315-321

[19] Lambin P, Debbia M, Béolet M, Brossard Y, Le Pennec PY, Rouger P. Quantitative estimation by ELISA of IgG anti-D (RH1) antibodies in immunoglobulin preparations and in the sera of immunized donors. Transfusion Clinique et Biologique. 2001;**8**(1):17-22

[20] Schäffner G, Kayser T, Tönjes A, Volkers P. Validation of flow cytometry to quantitate the potency of anti-D immunoglobulin preparations. Vox Sanguinis. 2003;**84**:129-136

[21] Hilden JO, Backteman K, Nilsson J, Ernerudh J. Flow cytometric quantitation of anti-D antibodies. Vox Sanguinis. 1997;**72**:172-176

[22] National Institute for Biological Standards and Control. 1st International Reference Preparation for anti-D Immunoglobulin, Human. NIBSC Code: 68/419. Instructions for Use, Version 2, 1 April 1998. South Mimms, UK: NIBSC; 1998

[23] Austin EB, McIntosh Y. anti-D quantitation by flow cytometry: A comparison of five methods. Transfusion. 2000;**40**:77-83

[24] Fox B, Roberts G, Sharp G, Studholme L, Atkinson E, Rigsby P, Malik K, Matejtschuk P. International Collaborative Study to Calibrate Proposed 3rd WHO International Standard for anti-D Immunoglobulin. WHO Expert Committee on Biological Standardization. WHO/BS/2018.2332 Geneva. 2018

[25] Duncan DB. *t*-Tests and intervals for comparisons suggested by the data. Biometrics. 1975;**31**:339-359

[26] Finney DJ. Statistical Methods in Biological Assay. 3rd ed. London: Charles Griffin; 1978

[27] Altman DG. Practical Statistics for Medical Research. 1st ed. London: Chapman & Hall; 1991

[28] Anonymous: Statistical analysis of results of biological assays and tests, 5.3. In: European Pharmacopoeia. 3rd ed. Strasbourg: European Commission; 2001

[29] Oviedo S, Romagnoli P, Collino C, Kassuha D, Vitali S. Desarrollo y Optimización de un Ensayo de Citometría de Flujo para Cuantificación de IgG-ANTI-D (Rho) en muestras del Proceso Productivo, (Reporte Corto). Revista Argentina de Hematología. 2001;**5**(2):117

[30] The International Conference on Harmonisation (1995): Note for guidance on validation of analytical procedures: Methodology (CPMP/ICH/281/95). European Agency for the Evaluation of Medicinal Products. 1996. Available from: http://www.eudra.org/emea.html

[31] WHO: 28th report. Expert Committee on Biological Standardization. WHO Technical Report Series. 1977;**610**:29-30

Section 3

Telocytes

Chapter 4

Telocytes: New Connecting Devices in the Stromal Space of Organs

Dragos Cretoiu, Maria Giuliana Vannucchi, Yihua Bei, Mirko Manetti, Maria Simonetta Faussone-Pellegrini, Lidia Ibba-Manneschi, Junjie Xiao and Sanda Maria Cretoiu

Abstract

Telocytes (TCs) represent a new type of interstitial cells, and were discovered by Prof. Popescu and his collaborators from Bucharest in 2005, and described as Interstitial Cajal-Like Cells (ICLCs). In 2010, Prof. Popescu and Prof. Faussone-Pellegrini from Florence, based on their expertise in morphology, agreed that in fact ICLCs were a brand-new entity and they renamed them telocytes. TCs are characterized by specific veil- or ribbon-like extensions called telopodes. Telopodes aid TCs in forming homo- or hetero-cellular contacts; thus, assembling three-dimensional networks that organizes the stromal and the parenchymal components of the organs. TCs can transfer information to neighbor cells ensuring a short-distance communication, and remotely by the release a wide variety of extracellular vesicles: exosomes, ectosomes, and multivesicular bodies. Here, we reviewed the evolution of the interest regarding TCs in different organs, in normal and pathological conditions. The main focus was on the role of TCs in gastrointestinal tract, urinary bladder, reproductive tract, and heart. This chapter sums up information about the possibilities that TCs are capable to behave as sensors/mediators in nervous activity, to represent mesenchymal stem cell precursors in adulthood, and to control and determine the differentiation/maturation of other cell types either during development or in postnatal life.

Keywords: telocytes, telopodes, podoms, podomers, 3D-network

1. Introduction: the telocytes

Discovered 13 years ago, telocytes (TCs) still represent a subject of debate regarding their role. TCs were first described in 2005, in the interstitial space surrounding exocrine acini in the pancreas, and considered as cells closely resembling the interstitial cells of Cajal (ICCs) by Romanian scientists from Carol Davila University of Medicine and Pharmacy in Bucharest, Romania, which entitled their first publication *Interstitial Cells of Cajal in Pancreas* [1]. Subsequent publications of this team described these cells under the name of interstitial Cajal-like cells and characterized them with the aid of electron microscopy and immunoctyo(histo) chemistry [2–6]. In 2010, professor Popescu, the Bucharest team leader, along with professor Faussone-Pellegrini, who is considered to be the international leading expert in ICCs (the gut pacemaker cells), have agreed that TCs and ICCs can be regarded as completely different cell populations based on the ultrastructural peculiarities, and went on to give them the name of "telocytes" [7]. In the following years, TCs have been described in numerous organs, and appear to be omnipresent in stromal space of humans and laboratory mammals where they form networks [8–10] by making contacts with each other (homo-cellular contacts). This is possible due to their unique and exceptionally long (several tens to hundreds of μm) cell prolongations called telopodes [7]. As professor Popescu liked to say "the shortest possible definition for telocytes is cells with telopodes" [11]. Telopodes are characterized by a succession of thin, filamentous regions (podomers) and dilated areas (podoms) with bead-like appearance [12]. TCs were revealed to have a third dimension that was only recently observed by FIB-SEM tomography, telopodes being represented by a veil- or ribbon-like extensions that compartmentalize the interstitial space [13]. Apart from the ability to form networks, TCs also can provide local communication with different cells and also to deliver extracellular vesicles to establish distant communication [14–18]. A detailed representation of the TCs contacts is illustrated in **Figure 1**.

To find out the function(s) of these cells, many methods of investigation have been employed, ranging from classical microscopy—optical and electronic—to advanced genomics and proteomics techniques. Thus, differentiation of TCs from mesenchymal stem cell adipocytes, fibroblasts, and endothelial cells was achieved [19–22].

The discovery and description of TCs have given rise to many controversies. From the publication of the first article to the present, the interest generated by TCs can be traced in the graph summarizing the number of articles published annually in PubMed (**Figure 2**). Though not yet described in speciality treatises as a distinct cellular type, this prospect does not seem to be far, the evidence being the increase in the number of articles published in the most prestigious journals such as Nature [23–25]. The discussion of these controversies is dealt with in detail by two recent reviews, in which there are enough arguments for and against the existence of TCs as a new cell type [26, 27]. We believe that in the shortest time new biomarkers will

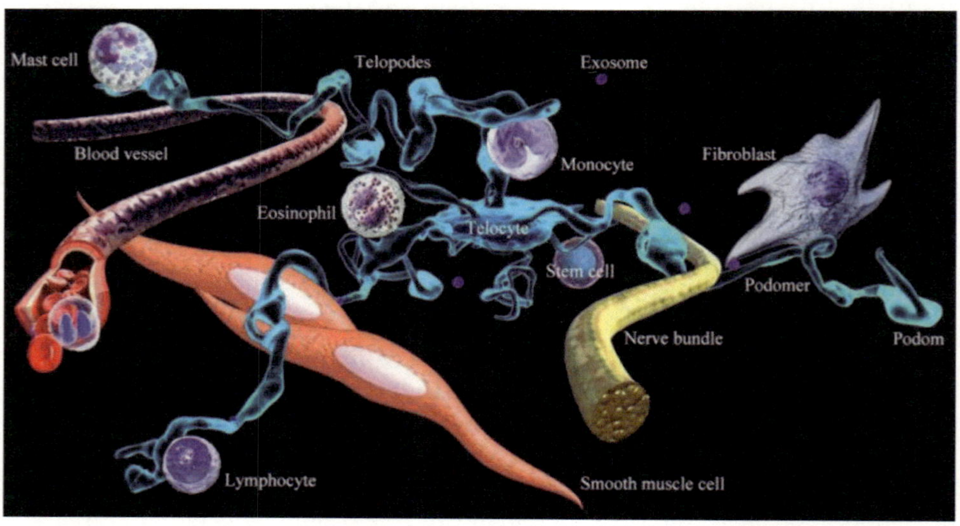

Figure 1.
Artistic representation of a 3D view of the contacts of a telocyte. TCs are regarded as interconnection devices due to their homo- and hetero-cellular junctions, as well as to their proximity to structures like blood vessels, nerve fibers, and muscle fibers. Image courtesy of Iurie Roatesi. Reproduced with permission from Ref. [75].

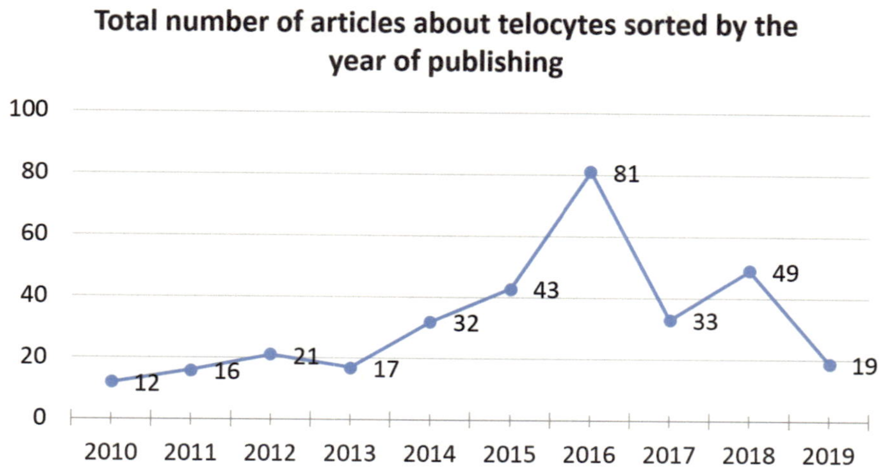

Figure 2.
Trends of publications searched in PubMed with "Telocytes" as a key word between 2010 and 2019.

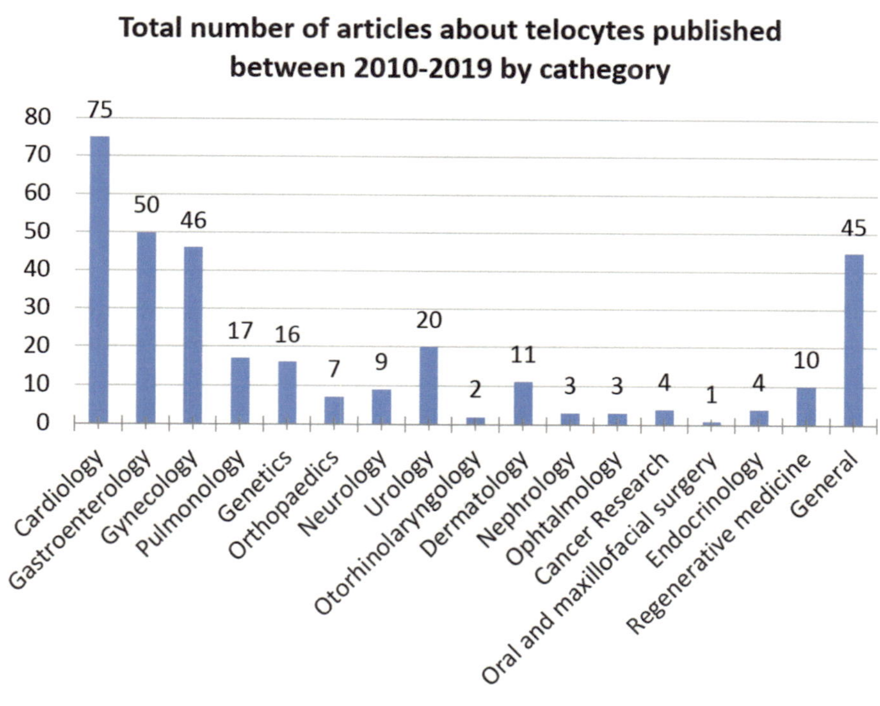

Figure 3.
Histogram of publications on telocytes categorized by categories of interest.

be found to prove the existence of this particular type of cell, and here is just one step to the description of their possible functions.

Our attention in this chapter is focused on synthesizing the available information relating to the main categories of interest that emerge from the chart shown in **Figure 3**, namely gastrointestinal tract, urinary bladder, reproductive tract, and heart.

1.1 The telocyte network

The ability of these cells to form 3-D stromal networks can be considered a discriminant element for their recognition under the light microscope [8], especially because no specific immunomarkers are available [28]. Indeed, the cell surface glycoprotein CD34, a marker shared with vascular endothelial cells, is currently considered one of the most suitable for the immunohistochemical identification of the TCs, which are also referred to as CD34+ stromal cells/TCs by some authors [29, 30]. Through extensive homo-cellular networks, TCs are believed to build the stromal scaffold whose continuity and adaptability guarantees the maintenance of the integrity of tissues/organs every time they are subjected to mechanical forces, such as distension and stretching. Moreover, TCs are universally considered key organizers of the connective tissue and eventually, they may contribute to the production and shaping of the extracellular matrix (ECM) in cooperation with fibroblasts. This has been observed in TCs located in the female genital tract where these cells express both estrogen and progesterone receptors [4, 31] whose activation is followed by significant changes of the TCs that acquire fibroblast-like features and become capable to produce the ECM [8]. Likely, the homo-cellular TCs contacts are also involved in the intercellular exchange of molecular or ionic signaling. Alongside the aforementioned roles, probably shared by all the TCs, many other roles have been attributed to these cells [30]. Therefore, each of the TCs subtypes is likely to play its organ-/tissue-specific role [8].

Although the TCs homo-cellular contacts are commonly observed, a variety of cell-to-cell contacts between TCs and other cell types (referred to as hetero-cellular contacts) are also observed [8, 32–35]. They consist of minute junctions (point contacts, nanocontacts, and planar contacts) whose mean inter-membrane distance is 10–30 nm, but more often by variably extended simple apposition of the contiguous plasma membranes that might act either as mechanical cell-to-cell attachments or as sites of intercellular communication [18]. Among these contacts, there are the so-called "stromal synapses" [36], a term used to describe those contacts occurring between TCs and several types of connective tissue cells such as mast cells, macrophages, myofibroblasts, and fibroblasts [8, 18, 35, 37]. The networks built by these hetero-cellular contacts are named "mixed networks." Collectively, the existence of mixed networks in addition to the homo-cellular TCs networks, the morphological and immunohistochemical differences reported for the TCs among organs and tissues, the existence of TCs subtypes, the interactions that TCs make with the ECM and, finally, the TCs vicinity to nerve endings and vascular cells, have substantiated the hypothesis that these cells may be part of integrated systems playing tissue-/organ-specific roles [8, 30, 32, 33, 35, 38].

1.2 Specific roles of the telocyte network: hollow organs

A common role proposed for the 3-D TCs scaffold in the hollow organs is to follow organ distension and relaxation avoiding anomalous deformation and controlling blood vessels closure or rheology. However, because of the anatomical complexity of such districts and the great variety of cell populations herein interacting with the TCs, many other roles are conceivable, suggesting that these cells, as connecting devices in the stromal space, might take center stage in the integration of all the information coming from the vascular, nervous, and immune systems, as well as from tissue-resident stem cells.

The present overview of the literature focused on the spatial organization, morphological, and histochemical peculiarities of the TCs, according to their location in different organs. This may help to point out the presumptive roles of the homo- and hetero-cellular TCs networks. With this aim, the TCs networks located in some representative hollow organs that have been more intensively studied, such as the gastrointestinal and reproductive tracts, the urinary bladder, and in the cardiac parenchyma, in both healthy and disease conditions, have been taken into consideration.

2. Gastrointestinal tract

The gastrointestinal tract consists of different hollow organs which have some similar and some different shapes and functions. Every organ modifies its lumen caliber and thickness several times throughout the day, following food transit. Food intake might happen several times per day, with different types and quantities, and the food transit varies according to the different regions, from the stomach to the colon. The cells of the lining epithelium do not change their shape, while microvilli height importantly changes. Under the mucosa, made by the epithelium, the lamina propria, and the muscular mucosae, there is the submucosa that has different morphological organization and function. Finally, the muscle coat is responsible for gastrointestinal contractility. Two motile activities, coordinated by the enteric nervous system and the ICCs, are present: peristalsis, a constant ab-oral movement that does not importantly modify the lumen caliber, and the relaxation/contraction related to food arrival/mixing for digestion and absorption/transit which promotes sustained lumen caliber changes. Region-specific mechanical and functional interrelationships between all the components of this complex apparatus are at the basis of the correct and coordinated behavior of the apparatus.

2.1 Gastrointestinal telocyte network in healthy condition

In the human and mouse gastrointestinal tract, the TCs form widespread networks in the mucosa, submucosa, muscle layers, at the myenteric plexus level, at the submucosal border of the muscular mucosae, in the circular muscle layer, and around nerve strands, blood vessels, funds of gastric glands, and intestinal crypts [32, 39]. Immunohistochemically, all the TCs residing in the different layers of the gastrointestinal tract wall can be identified as CD34+/PDGFRα+ interstitial cells [32]. In the lamina propria and submucosa, the 3D homo-cellular TCs network has a structural role, forming the scaffolding that can direct the collagen fibers/bundles and define the spaces where the several elements of the connective tissue accommodate. Although it cannot be excluded that these TCs could eventually be recruited for the ECM synthesis, the abovementioned structural function is likely to be the main one. However, the role attributed to the TCs lining the basal-lateral surface of the glandular crypts [32, 39], where epithelial stem cells are located is particularly intriguing, since these TCs have been proposed to influence the proliferation and differentiation of stem cells due to their ability to produce and secrete a variety of molecules [40], the close relationships they recurrently establish with the "stem cell niches" [34, 40], and the expression on their surface of the functional receptor PDGFRα whose activation is critical in mammalian organogenesis [39]. In this context, it is worth mentioning that a very recent study demonstrated that the subepithelial plexus formed by PDGFRα+ TCs acts as a crucial source of Wnt proteins, which are essential to support intestinal crypt stem cell proliferation and epithelial renewal [23]. In the muscle coat, by both immunohistochemistry (PDGFRα and CD34 immunolabeling) and electron microscopy, the TCs processes were observed to constitute 3D networks intermingling with those of the ICCs and to establish cell-to-cell contacts with them [32, 34]. Interestingly, within the gut muscle layers the TCs and ICCs networks can be clearly distinguished based on their different immunophenotypes, as the TCs are CD34+/PDGFRα+ and negative for c-kit, and vice versa, the ICC are c-kit+ and negative for either CD34 or PDGFRα [32]

(**Figure 4A–I**). This mixed TCs/ICCs meshwork and areas of simple apposition occurring between the TCs and the smooth muscle cells have suggested that the intramuscular TCs might support the spreading of the slow waves generated by the ICCs, which are electrically coupled to the smooth muscle cells, thus contributing to the regulation of gastrointestinal motility [32, 41]. In favor of this hypothesis, it has been recently reported that the "smooth muscle cells are electrically coupled to both ICCs and PDGFRα+ cells (i.e., the TCs) forming an integrated unit called the SIP syncytium" [42]. Another possible role attributed to the TCs located in the gut muscle coat is that they might eventually differentiate into ICCs. This hypothesis is mainly based on the existence of the ICCs/TCs mixed network, where the two interstitial cell types are often intercalated (**Figure 4I**) [32, 34, 37]. In support of this hypothesis, despite apoptotic ICCs have been described in the colon of human healthy subjects of different ages, no decrease in the number of ICCs was observed in relation to aging and no ICCs was ever seen undergoing mitosis [43], while mitotic TCs rich in rough endoplasmic reticulum can be detected in the interstitial spaces usually occupied by the ICCs (personal unpublished observation). Taken together, these data suggest that TCs might represent a pool of ICCs precursors being responsible for the physiological replacement of aged ICCs. Furthermore, it has been demonstrated that in culture stromal cells expressing CD34 (a typical marker of the TCs) proliferate and progressively lose their CD34-positivity to acquire the c-kit-positivity (a typical marker of the ICCs) [44]. Reasonably, in adulthood the TCs, wherever they are located, might be considered as a pool of mesenchymal stromal cells and, in the gut, to be important for ICCs renewal [37].

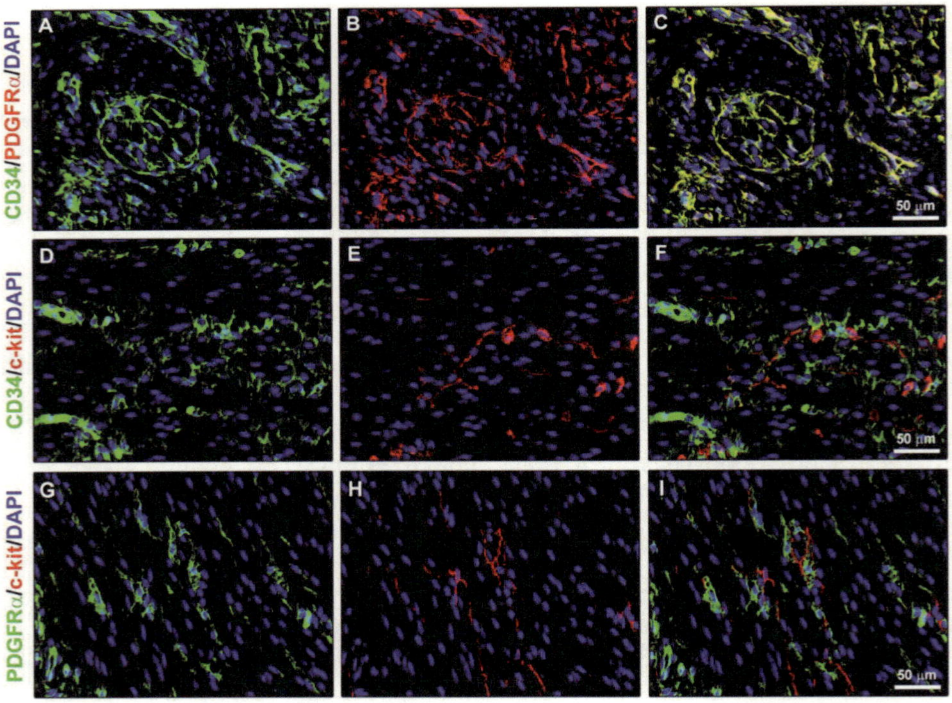

Figure 4.
TCs and ICCs in the human gastrointestinal tract. TCs and ICCs form intermingled networks in the muscularis propria of the human intestine. Representative images of human colon sections double immune-stained for: (A–C) CD34 (green) and PDGFRα (red), (D–F) CD34 (green) and c-kit (red), and (G–I) PDGFRα (green) and c-kit (red) are shown. Nuclei are counterstained with DAPI (blue). Merge images are shown in the right panels. All the TCs are CD34+/PDGFRα+ (A–C), while the ICCs are c-kit+ and negative for either CD34 (D–F) or PDGFRα (G–I). Scale bar: 50 μm.

2.2 Telocyte network in gastrointestinal diseases

Inflammatory bowel diseases (IBD), including Crohn's disease (CD) and ulcerative colitis (UC), are complex disorders in which chronic relapsing and inflammation progressively evolve into extensive fibrosis of the intestinal wall [41, 45]. Both CD and UC are characterized by abdominal pain and diarrhea, mainly as a result of the progressive fibrotic process that leads to a stiff intestine unable to properly carry out peristalsis and resorptive functions [45–47]. The evidence that intestinal dysmotility with a reduction in the number of ICCs is a feature of IBD and that the ICCs and TCs networks are intermingled in the gut neuromuscular compartment has prompted an investigation of the TCs distribution either in the terminal ileum of CD patients or in the colon of UC patients [32, 41, 48–51]. Interestingly, in both conditions, the gut wall fibrosis was strictly paralleled by a reduction in TCs [50, 51]. In fact, in the CD intestinal wall, which is histopathologically characterized by discontinuous signs of inflammation and fibrosis (referred to as "skip lesions"), TCs were normally distributed in all layers of the healthy-looking areas from the mucosa to the subserosa, while they were markedly reduced in the fibrotic areas displaying severe architectural derangement [50]. In particular, the network of TCs was discontinuous or even completely lost among smooth muscle bundles and around myenteric plexus ganglia [50]. As far as UC is concerned, TCs were investigated in tissue specimens from both patients in early and those in advanced phases of colonic wall fibrotic remodeling [51]. In the early phase, fibrosis is confined to the muscular mucosae and submucosa, while in the advanced one it extends affecting wide areas of muscle layers and the myenteric plexus. Of note, TCs were significantly reduced in the muscular mucosae and submucosa of both early and advanced fibrotic UC cases [51]. On the contrary, the intramuscular and myenteric plexus TCs networks were severely compromised in the advanced but not in the early fibrotic UC cases [51]. Through double immunofluorescence, it was possible to further reveal that in both forms of IBD the losses of TCs and ICCs occurred in parallel in the muscle layers and around the myenteric ganglia [50, 51]. Based on these findings, it has been proposed that the simultaneous reductions in the TCs and ICCs might significantly account for intestinal dysmotility in IBD. Several assumptions have also been made concerning the possible causes and pathophysiologic implications of this TCs impairment [41, 50, 51]. As reported in the failing human heart [41, 52], the progressive alteration in the ECM composition and entrapment of TCs in such fibrotic ECM may provoke profound cell sufferance and eventually lead to cell death. Both the ECM accumulation/rearrangement and the parallel reduction of TCs may profoundly impair their hetero-cellular networks with immune cells, fibroblasts, smooth muscle cells, ICCs, blood vessels, and nerve endings, thus hampering the TCs intercellular signaling functions [41]. Whether the loss of the TCs might even precede the onset of fibrosis rather than being merely a consequence of tissue fibrotic remodeling is difficult to demonstrate. In line with the proposed role of TCs as a guide for the correct tissue shaping during organ morphogenesis, it cannot be excluded that the loss of TCs might contribute to the altered 3D ECM organization in the fibrotic intestinal wall [46]. For instance, it has been proposed that the disappearance of TCs might favor an uncontrolled activation of ECM synthesizing fibroblasts and their transition to profibrotic α-smooth muscle actin (α-SMA)+ myofibroblasts [46]. Noteworthy, hetero-cellular contacts between TCs and fibroblasts/myofibroblasts have been described in different organs, suggesting that the TCs could contribute to tissue homeostasis by controlling the synthetic activity of such partners through inhibitory signals [46]. In support of this hypothesis, in

UC colonic specimens the loss of TCs was paralleled by an increase in the number of α-SMA+ myofibroblasts [51]. This last observation suggests that during pathologic processes a subset of TC could undergo phenotypic changing, possibly contributing to the increase in the profibrotic myofibroblast population [41]. However, double immunolabeling for CD34 (as TC marker) and α-SMA did not reveal the presence of CD34+/α-SMA+ transitioning stromal cells in the colonic wall of UC patients, which makes unlikely the aforementioned hypothesis [51]. Also, electron microscopy investigations of different pathological tissues (e.g., failing human heart, fibrotic skin) have clearly shown that fibrosis is accompanied by TC degenerative processes rather than by activation/transformation into myofibroblasts [52, 53]. As the findings in the CD and UC intestinal tissues, a loss of TCs have also been reported in the neuromuscular compartment of the fibrotic gastric wall of patients with systemic sclerosis, where it likely contributes to gastric dysmotility clinically manifesting as delayed gastric emptying or gastroparesis [54].

Besides IBD, recent evidence suggests that the TCs might be a cell source of certain kinds of gastrointestinal stromal tumors (GIST) [55]. While ICCs hyperplasia has been identified as a crucial pathogenic feature of KIT-mutant GIST, it has been proposed that the TCs could represent the physiological counterpart of PDGFRa-mutant GIST and inflammatory fibroid polyps [55]. Indeed, a pathogenic relationship between TCs hyperplasia and both inflammatory fibroid polyps and PDGFRa-mutant GIST has been suggested. Moreover, the term "telocytoma" was proposed for defining inflammatory fibroid polyps, since it conveys both the pathogenic (neoplastic) and histotypic ("telocytary") nature of this tumor [55].

3. Human urinary bladder

The urinary bladder is a complex organ which modifies its volume and shapes several times during the day following filling and micturition. Filling happens gradually during the time while micturition happens in a unique and sustained emission of the urine. The cells of the lining epithelium (urothelium) dramatically changes their shape and height; under the urothelium, there are two regions: the upper lamina propria (ULP) and the deep lamina propria (DLP), with different morphological organization and function; the detrusor is the muscle coat responsible for organ relaxation and contraction. Region-specific mechanical and functional interrelationships between all the urinary bladder components are at the basis of correct and coordinated behavior.

3.1 Bladder telocyte network in healthy condition

In the bladder, unlike in other organs, a more complex picture comes out [33, 35]. As is the case in the gut, the bladder TCs form complex networks and make contacts either between themselves or with other cell types; however, depending on their location, they show different immunohistochemical properties and ultrastructural peculiarities. The main differences are detected between the TCs located in the sub-urothelial connective tissue (upper lamina propria, ULP) and those located in the submucosa (also referred to as deep lamina propria) and detrusor. The TCs in the ULP is PDGFRα+ and CD34- while those in the submucosa and detrusor are CD34+ and PDGFRα- (**Figure 5A** and **B**). Moreover, while the TCs immediately below the urothelium express only the PDGFRα and form a homo-cellular network, the other sub-urothelial TCs are also α-SMA+ (**Figure 5A**)

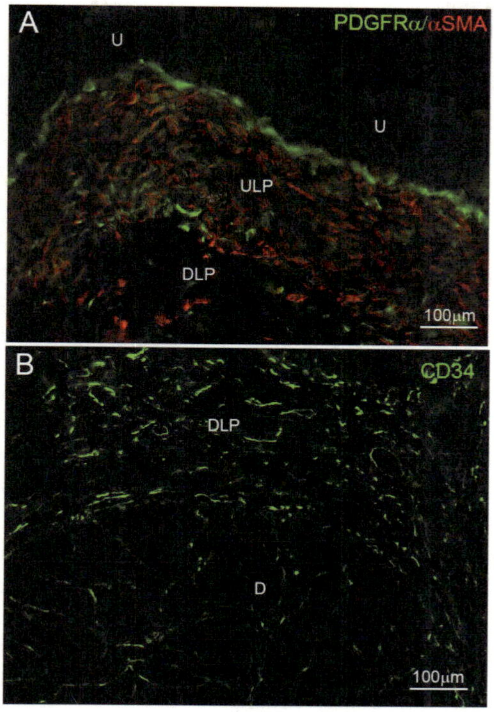

Figure 5.
TCs networks in the human urinary bladder. (A) PDGFRα (green) and α-SMA (red) immunolabeling. A monolayer of PDGFRα+ TC lines the urothelium (U). In the remaining upper portion of the lamina propria (ULP) a mixed network made by PDGFRα+/α-SMA+ TC (hybrid TC) and α-SMA+ myofibroblasts is present. (B) CD34+ TC form a homo-cellular network in the deep lamina propria (DLP) and detrusor (D). Scale bar: A, B = 100 μm.

and under the transmission electron microscope show a larger body and cell processes possessing attachment plaques with the connective tissue like the fibronexuses typical of myofibroblasts. Further, these TCs establish extended regions of simple apposition with the myofibroblasts, thus forming a 3D mixed network (**Figure 5A**). This mixed network together with the homo-cellular networks that the TCs make in the remaining portions of the bladder constitutes the scaffolding to guarantee the organ integrity during distention and relaxation [33, 35, 37, 56]. However, other specific roles have been assumed for the bladder TCs, particularly for those located in the ULP. The TCs lining the urothelium is quite peculiar, because of their location and immunolabeling [33, 38]. As already discussed, the PDGFRα+ TCs located immediately beneath the intestinal crypts appear to be cells engaged in controlling the proliferation and differentiation of the stem cells resident in the crypts [23]. Likewise, sub-urothelial TCs could play similar functions. Immediately below those TCs, there is the TCs/myofibroblast network and both cell types form gap junctions and express the Cx43 protein, are close to nerve varicosities, express the vanilloid, the ATP, the purine and the muscarinic receptors and contain the cGMP, the target molecule of NO [35]. These features support the hypothesis that these cells have a role as intermediaries in propagating chemical or electrical stimuli locally generated, as well as a role as the target of the paracrine activity of the urothelium and the nervous stimuli [35, 38, 57]. The importance of these roles is plenty understood, since the ULP and the urothelium constitute a sensory system capable of perceiving mechanical and chemical stimuli and whose integrated responses control the efferent pathways on the detrusor and the micturition.

3.2 Telocytes in bladder diseases

The micturition reflex is the result of a complex integration among involuntary and voluntary nervous mechanisms. Several pathological conditions of the low urinary tract compromise this function causing detrusor dysfunctionality. The most frequent is the idiopathic detrusor overactivity, the neurogenic detrusor overactivity (NDO), the bladder pain syndrome/interstitial cystitis, and the partial bladder outlet obstruction. All these diseases are characterized functionally by excessive sensitivity of the detrusor/bladder to filling [58], and histologically by an intense inflammation especially in the lamina propria [35, 38, 59, 60]. Since it is well known that TCs and myofibroblasts produce cytokines and other molecules able to recruit immune cells and express receptors for the cytokines released by the immune cells, both cell types likely intervene in the inflammation intensity, quality, and duration. Furthermore, in the presence of detrusor hyperactivity, both ULP-TC and myofibroblasts showed an increase of the Cx43 protein labeling that was interpreted as an augmentation of the gap junctions and signs of cellular activation (clear nuclei and larger bodies) [35, 38, 59, 61]. Additionally, the TCs expressing both PDGFRα and α-SMA were significantly increased in comparison with controls suggesting a shift toward a myofibroblast phenotype [35, 38]. All these cell changes were considered signs of adaptability because, despite the presence of inflammation, the 3D cell network was preserved [38, 59]. However, this integrity not necessarily means adequate functionality of the sensory system made by the urothelium and the ULP; in fact, the higher thickness of the ULP, due to the intense cell infiltrate and edema, forcing the net meshes to enlarge, could cause an increase in the distances among the cells, between them and the nerve endings, and between all of them and the urothelium, likely affecting the sensitivity to volume changes and the capability of responding to the molecules released by the nerve terminals and by the urothelium. Finally, because the ULP thickening was uneven alongside the organ [35, 38, 60], foci of hypersensitivity could alternate to others less responding, further prejudicing the correct integration of the afferent stimuli [57]. Finally, it was reported that the TCs forming the monolayer underlying the urothelium did not show any significant changes in hyperactive bladders. These data were explained as follow: the location of these TCs could spare them from the damages caused by the cell infiltrate. Further, if these TCs are engaged in cell proliferation and differentiation of the overlying epithelium, the absence of epithelial cell death signs in NDO might account for their sparing.

4. Reproductive system

The female reproductive system includes, besides external sex organs, the internal sex organs: the ovaries, fallopian tubes, and uterus. Immature at birth, these organs continue to develop and reach maturity at puberty when they can produce gametes, and to carry a fetus to full term. Fallopian tubes integrity is capital for fertilization which usually occurs in the external third of the tubes. The traveling zygote will form the blastocyst that will be implanted in the uterine endometrium. To obtain and maintain a pregnancy, the integrity and functionality of these organs need to be at a maximum.

4.1 Telocytes in uterus and fallopian tubes in healthy condition

Currently, TCs are found in uterine tubes and uterus, including endometrium, myometrium, and cervix, and also in the vagina [2, 3, 62–65]. Among the first locations in which TCs were described are the organs of the female genital apparatus: the uterus and the fallopian tubes [2, 3, 5, 10, 66]. Since the beginning, their characterization was based on conventional microscopy methods and techniques such as methylene blue staining and silver impregnation, in situ and in vitro [2, 3], followed by the description of the "gold standard" for their identification with the aid of electron microscopy [5]. In parallel, various immunohistochemical markers have been used to enhance TCs better characterization, which has varied over time from vimentin, α-SMA, progesterone receptor, desmin, estrogen receptor, and S100 protein, to stabilize what we consider nowadays to best describe the phenotype of these cells—CD34 and PDGFRα [2, 3, 67]. A book chapter refers to the immunohistochemistry of TCs in female genital organs [28]. However, it should be pointed out that TCs in the uterus and fallopian tubes express receptors for estrogen and progesterone [4, 31]. Nowadays, the most suitable methods for TCs identification are electron microscopy and double staining for CD34 and PDGFRα or PDGFRβ or vimentin [68, 69].

In the human uterus, TCs establish homo- and hetero-cellular junctions, demonstrated by electron microscopy images of the telopodes (**Figure 6**) [70]. Homo-cellular junctions are typically established between two telopodes, but might be observed between a telopode and a TCs' body [71]. The most frequently observed interactions are simple appositions of the plasma membranes; however, puncta adhaerentia minima, processus adhaerens, recessus adhaerens, and manubria adhaerentia can be captured in electron microscopy images [10, 18]. Sometimes, even gap junctions were captured [18]. Hetero-cellular junctions are usually seen between telopodes and fibroblasts, myofibroblasts, pericytes, stem cells, macrophages, mast cells, lymphocytes, plasma cells, Schwann cells, endothelial cell, neurons, cardiomyocytes, and smooth muscle cells, as described in the literature [12, 56, 72–74]. The contacts made between the two membranes are of the type of point contacts, nanocontacts, planar contacts, or simple apposition of plasma membranes [18].

TCs can release exosomes (from multivesicular bodies), ectosomes (shredded directly from plasma membrane), and multivesicular cargos (multiple tightly packed endomembrane-derived vesicles) [75]. The three types of extracellular vesicles emitted by TCs are evidence of the involvement of these cells in intercellular distance communication. Shed vesicle number and diameter are not correlated with the reproductive state, while the quantity of TCs in the endometrium and the myometrium varies with it [17]. Moreover, it was demonstrated that the morphology of telopodes is correlated with the presence or absence of gestation [70].

All these morphological, immunohistochemical, and electrophysiological observations have led to several hypotheses on the TC functions in the uterus and fallopian tubes. The existence of homo-cellular junctions leads to a presumptive function in controlling the shape of the tissues which are subjected to of dynamic changes, such as the pregnant uterus that hypertrophies and expands as the fetus grows [17]. In support of this assumption stands the hypothesis that TCs contribute to smooth muscle growth in areas with high mechanical forces due to TCs mechanical sensitivity [76]. The mechano-sensing function should also be considered, due to the presence of catenins that make up the junctions [18, 71]. Moreover, TCs express T-type calcium (CaV3.1 and CaV3.2) channels and small-conductance calcium-activated potassium channels (SK3) and calcium-dependent hyperpolarization-activated chloride inward channels, the levels of expression being dependent on the physiological state, pregnant or non-pregnant [70, 77, 78]. This point to a TCs' involvement in calcium signaling mechanisms with neighboring cells [79]. Extracellular matrix remodeling is also emphasized by some studies and can be considered as applicable to the uterus [80, 81]. The existence of (ERα)

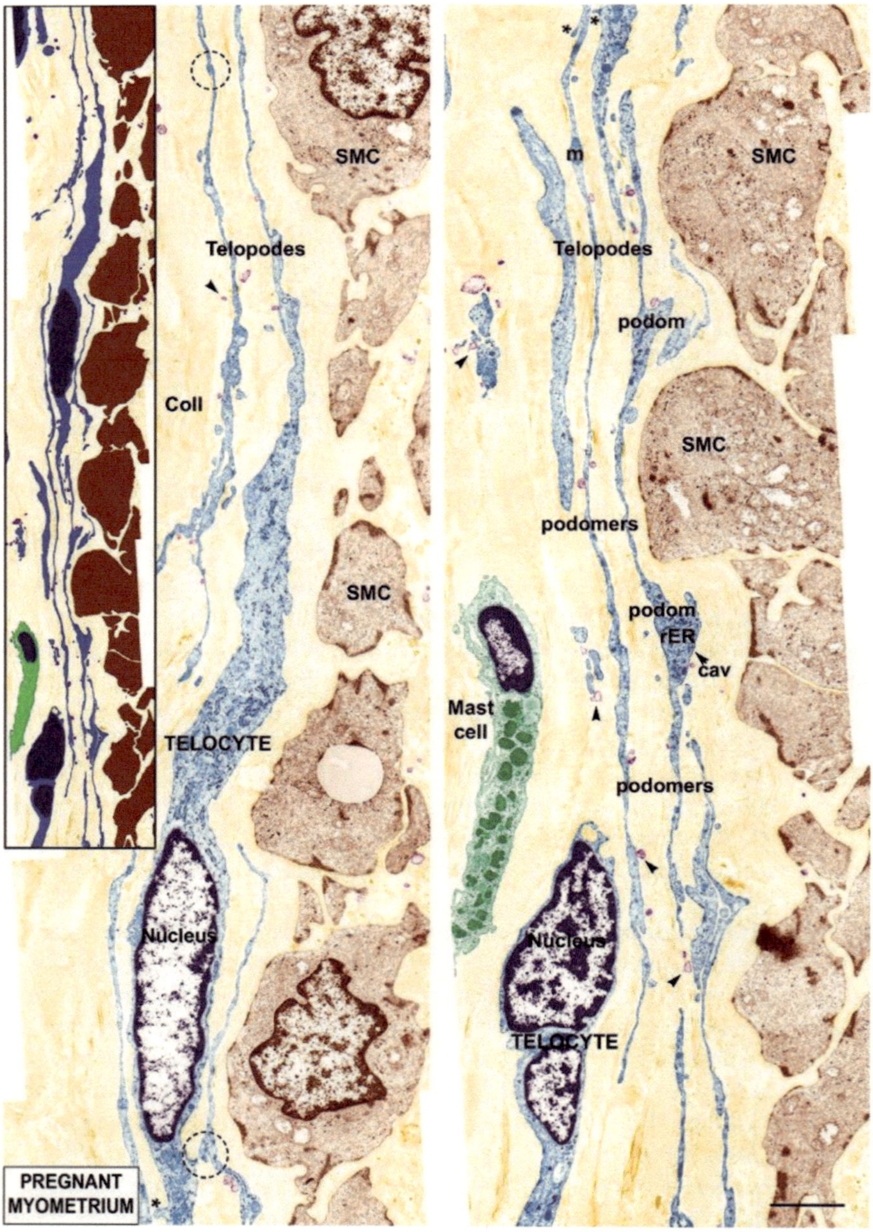

Figure 6.
Representative ultrathin section of human pregnant myometrium. Two-dimensional sequenced concatenation from 11 serial electron micrographs showing the 3D network of TCs (blue) interconnected by homo-cellular junctions (dotted circles). SMCs are shown in cross-section and were digitally colored brown. In their vicinity, numerous Tps (blue) establish a network and release extracellular organelles (exosomes and shedding vesicles [arrowheads]) digitally colored purple. One mast cell (green) is in the vicinity of this network. Some vesicles were captured at the moment of being shed from Tps (*). Cav = caveolae; coll = collagen; m = mitochondria; rER = rough endoplasmic reticulum; N¼nucleus. Bar¼2 lm. Reproduced with permission from Ref. [10].

and progesterone receptor A (PR-A) on the surface of uterine TCs suggests their involvement as sensors for steroid hormones levels. Although little is known about the existence of stem cells in the uterus, they certainly exist, and the secretome of the TCs could influence the cellular microenvironment, controlling their proliferation, and differentiation [15]. Also, TCs secretome factors could participate in decidua formation [17]. Some studies pointed to the angiogenic properties of TCs due to vascular endothelial growth factor (VEGF) expression [15, 82, 83], while others indicated the anti-oxidative properties of TCs because the TCs specific morphology can be changed by modifying the redox balance of their environment or by aging due to their richness in SOD2 (mitochondrial superoxide dismutase) [21, 84].

Recently, TCs were found to activate and "educate" peritoneal macrophages (pMACs) with the aid of telopodes by direct physical contact through heterocellular junctions or using a TCs-conditioned media through paracrine mechanisms [85]. This is suggestive for a role in immunosurveillance [85].

In the fallopian tubes, TCs have been described throughout the thickness of the wall, their density decreasing from the mucosa to the muscular, from ~18 to ~7.5% [3]. TCs are also found in the fimbriae of Fallopian tubes [17, 28]. A panel of antibodies was used to identify tubal TCs. The telopodes possess all the features described above, creating a 3D network and establishing contacts with different structures, such as blood vessels, nerves, and muscle fibers [86, 87]. Both, homo- and hetero-cellular junctions are described, and also, a new hypothesis was emitted that telopodes that contact the immune cells (plasma cells and lymphocytes) might stimulate antibody production [88, 89]. As suggested by Cretoiu et al., the tubal peristalsis might suffer influences from TCs which also express PR-A and ERα receptors [31]. The tubal movements seem to be amplified by estrogen and decelerated by progesterone [90]. Recently, additional markers were tested for TCs identification, such as Podoplanin (D2–40) and Dog-1 but proved to be inappropriate [91].

4.2 Telocytes in uterine and tubal diseases

The pathogenesis of uterine leiomyomas, the most frequent benign tumors in women might be determined, among other factors, by the loss of TCs [26]. Varga et al. proposed three hypotheses regarding TCs' involvement: (i) loss of TCs as steroid sensors leads to an increased density of estrogen receptors at smooth muscle cells level followed by a cell cycle disruption; (ii) considered as progenitor cells, TCs absence can favor the rise of new leiomyoma cells; and (iii) in the absence of antioxidant protection conferred by TCs, leiomyoma cells grow numerically due to local hypoxia that blocks their apoptosis [26].

A recent study shows, for the first time, that there is an interplay between telocytes and autonomic innervation in leiomyomata [92]. TCs decreased in numbers in the leiomyomatous myometrium, suggesting a role for these cells in the control of the microenvironment [92].

The integrity of the 3D network of TCs appears to be fundamental in exercising the function of the fallopian tubes that are regarded as a major organ in the reproduction. Several studies showed that by affecting the 3D organization and number of TCs perturbations occur in the local homeostasis, leading to angiogenesis and interstitial fibrosis [88, 89]. Neo-angiogenesis plays a major role in endometriosis and adenomyosis pathogenesis, and even in tubal ectopic pregnancy [93–95]. TCs were shown to be involved in all these processes [96, 97]. In pelvic endometriosis and tubal ectopic pregnancy, the decrease in the number of TCs is probably due to the overproduction of iNOS, COX-2, LPO, and estradiol [49, 69, 88].

Some other pathologies that might affect the 3-D network of TCs were described, such as Chlamydia infection responsible for the activation of macrophages or pelvic inflammatory disease [89, 98, 99]. In inflammation and ischemia, TCs were shown to be lost and to suffer major ultrastructural changes, a process followed by interstitial fibrotic remodeling [99]. Abd-Elhafeez et al. proposed a role for tubal TCs in the regulation of the epithelial function necessary for the final gamete maturation, fertilization, and early embryo development [100].

5. Heart

Cardiac TCs are among the most well described in the body [101]. Cardiac TCs are unique interstitial cells in the heart [101]. Intramyocardial TCs account for less than 1% of interstitial cells in the human heart [56]. Cardiac TCs have been identified in heart valves, left and right atrium and ventricle, epicardium, myocardium, endocardium, sub-endocardium, and myocardial sleeves [101, 102], in mice, rat, porcine, and human [56, 103]. The whole ultrastructural anatomy of human cardiac TCs has been reconstructed by focused ion beam scanning electron microscopy (FIB-SEM) [104]. The electrophysiology of human cardiac atrial and ventricular TCs has also been reported [105].

Cardiac TCs are completely different from other types of myocardial interstitial cells, especially from cardiac fibroblasts [106]. Cardiac TCs and cardiac fibroblasts are completely different in immunophenotypes, as cardiac TCs are positive for CD34/PDGFR-α, CD34/PDGFR-ß, or CD34/Vimentin while cardiac fibroblasts are only positive for PDGFR-ß and Vimentin [106]. CD34/PDGFR-α positive TCs account for one-third of the total cells among TCs enriched rat cardiac interstitial cell population [106, 107]. Some studies have also reported that cardiac TCs inconstantly express CD34/c-kit [56]. Besides, cardiac TCs are also distinct from pericytes, since cardiac TCs are CD34 positive and α-SMA weak positive while pericytes are CD34 negative and α-SMA positive [106]. Moreover, cardiac TCs are CD34/PDGFR-ß positive while pericytes are CD34 negative and PDGFR-ß positive [106]. Interestingly, cardiac TCs are positive for CD29 (a mesenchymal marker) but negative for CD45 (a hematopoietic marker), suggesting that cardiac TCs could be a source of cardiac mesenchymal cells [106]. Interestingly, the telomerase concentration in CD117 and CD34 positive cardiac TCs is significantly higher than that in cardiomyocytes and is 2.5-times and 1.5-times lower than that in bone mesenchymal stem cells and cardiac fibroblasts [108].

Cardiac TCs can form tight junctions with all other types of cells within the heart including cardiomyocytes, vascular smooth muscle cells, endothelial cells, and pericytes. The functions of cardiac TCs are not fully known but are proposed as follows: (1) intercellular signaling; (2) mechanoreceptors/transducers; and (3) cardiac homeostasis and repair [101, 109]. In the heart, telocytes participate in cardiac development and physiology, and diverse cardiovascular diseases (**Figure 7**).

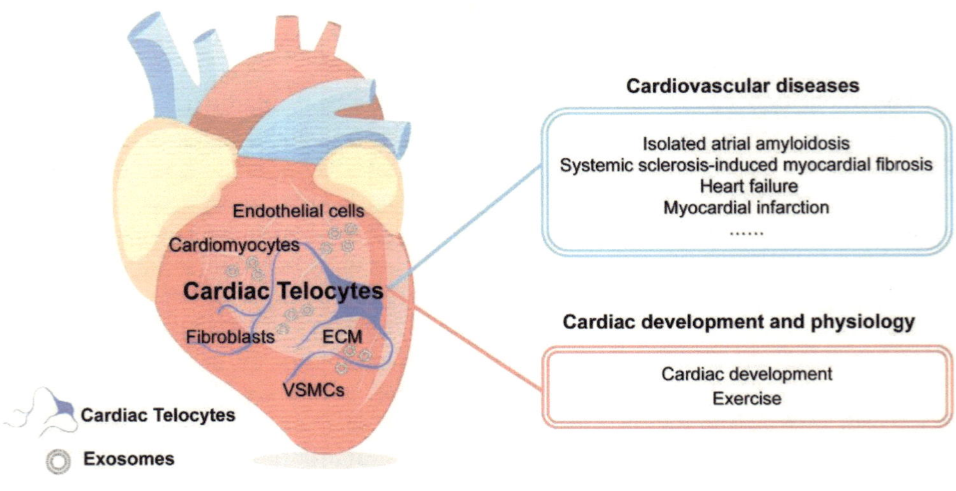

Figure 7.
Cardiac telocytes are in tandem with different types of cells in the heart, and participate in cardiac development and physiology, and diverse cardiovascular diseases. ECM, extracellular matrix; VSMCs, vascular smooth muscle cells.

5.1 Cardiac telocytes in heart development and physiological growth

Involvement of cardiac TCs and cardiomyocytes during development have been investigated using myocardium from embryonic (E14, E17), newborn (P0, P6), and adult (2 months) CD1 mice by using transmission electron microscopy and immunohistochemistry [110]. It was found that TCs were present from early embryonic to adult life in the mouse heart [110]. Besides, cardiac TCs demonstrated immature features in early embryonic hearts while cardiac TCs exhibit a more differentiated phenotype in newborn hearts [110]. Cardiac TCs played a fundamental role during cardiac development by forming a correct three-dimensional myocardial architecture and nursing cardiomyocyte precursors [110]. Intriguingly, cardiac TCs were found negative for c-kit and CD34 during the embryonic stage [110]; however, CD34 was expressed in a few TCs in the heart of a newborn mouse, and in most TCs in adult hearts [110]. This suggests a phenotype switch of cardiac TCs during development.

Exercise can induce cardiac physiological growth, which is characterized by increased cell size of cardiomyocytes and the formation of new cardiomyocytes [111, 112]. Three double-immunostainings, including CD34/PDGFR-α, CD34/PDGFR-ß-, and CD34/vimentin, have been used to determine the number of cardiac TCs in exercise-induced physiological cardiac growth [81]. The number of cardiac TCs was found to be significantly increased in the exercised heart [81]. The increased cardiac TCs in exercise might communicate with cardiomyocytes through direct contacts or telocyte-shed vesicles, balance angiogenesis, and maintain the normal 3D-organization of ECM [81]. This study suggested a potential role of cardiac TCs in exercise-induced cardiac growth [81].

5.2 Cardiac telocytes in cardiovascular diseases

Isolated atrial amyloidosis (IAA) is frequently found in long-standing atrial fibrillation patients [113]. By electron microscopy, telopodes are found surrounding the amyloid deposits, which limit their spreading into the interstitium [113]. This indicates that TCs might participate in amyloidogenesis by gathering masses of amyloid fibrils.

Systemic sclerosis represents a complex connective tissue disease featured with fibrosis of the skin and various internal organs [54]. TCs, as defined by CD34-positivity/CD31-negativity, were checked in the fibrotic areas of systemic sclerosis myocardium and were found to be almost undetectable [54]. However, in the control myocardium, numerous TCs were found located in the interstitium surrounding cardiomyocytes [54]. This indicates that loss of cardiac TCs contributes to myocardial fibrosis caused by systemic sclerosis.

The imbalance between cardiac TCs apoptotic death and cardiac TCs proliferation is responsible for the depletion of cardiac TCs in cardiac diseases leading to heart failure [52]. In human heart failure patients' myocardium, the number of cardiac TCs and telopodes decreases over twofold. Additionally, the apoptotic cardiac TCs increases threefold in the diseased heart while the percentage of proliferating cardiac TCs remains unchanged, suggesting that the decreased cardiac TCs population in heart failure is mainly due to increased apoptosis [52]. Interestingly, the number of cardiac TCs and telopodes has been found to depend on the composition of the extracellular matrix which is correlated negatively with mature fibrillar collagens and positively with degraded collagens [52].

The changes of cardiac TCs have been determined in an acute myocardial infarction rat model induced by isoproterenol (ISO) [114]. It was found that CD117/CD34 positive cardiac TCs were undetectable with immunohistochemical staining 1 day after ISO treatment. Interestingly, treatment with grape seed extract (GSE) could significantly increase cardiac TCs numbers and enhance angiogenesis in myocardia but not in other tissues; in fact, it was found suppressing angiogenesis in tumor tissues instead [114]. Thus, GSE was regarded to promote angiogenesis by modulating cardiac TCs, which subsequently stimulated endothelial cells [114].

Also, in a rat myocardial infarction model induced by coronary occlusion, cardiac TCs were reported undetectable in the infarction zone from 4 days to 4 weeks [115]. Simultaneous transplantation of cardiac TCs could significantly decrease infarct size and improve heart function 2 weeks after myocardial infarction [115]. Moreover, the protective effects of intramyocardial transplantation of cardiac TCs were also observed 14 weeks after myocardial infarction as evidenced by improved heart function, decreased infarct size, increased angiogenesis, and decreased myocardial fibrosis [116].

Currently, no single specific immunophenotype for cardiac TCs has been identified [101]. For in-depth studies of cardiac TCs, it is highly needed to identify a specific immunostaining marker for them. Most isolated cardiac TCs are either not pure enough (as cardiac fibroblasts grow much faster than cardiac TCs) or only containing subtypes of cardiac TCs. It would be beneficial to investigate the therapeutic effects of cardiac TCs or cardiac TCs-derived exosomes. Moreover, the immunoregulatory effects of cardiac TCs should be thoroughly investigated. Finally, other organs or tissue-specific TCs are worthy to be studied.

6. Conclusion and future overlook for telocytes' study

In conclusion, this review of the literature data indicates that TCs, depending on their location, may display different immunohistochemical properties, ultrastructural peculiarities and form complex networks making contacts either between themselves or with other cell types. Further, our current knowledge of TCs allows the following conclusions on the role(s) that these cells may play, some of which might be common to the different organs and some be organ-/region-specific.

1. In the stromal space of all the organs taken into account, TCs appear as connecting cells. Reasonably, TCs, due to their homo- and hetero-cellular contacts, can be considered as *connecting devices* playing either common or region-specific roles. These contacts might be merely mechanical or sites of cross-talking between TCs and other cell types establishing intercellular molecular exchanges. Spatial relationships also suggest an involvement of the TCs network in the coordination of tissue homeostasis in response to local functional demands.

The involvement in tissue homeostasis might be explained by the heterogeneity of TCs depending on their location.

2. TCs might be engaged in *controlling the proliferation and differentiation of the stem cells* and, either in the adulthood or during organ differentiation, wherever they are located, these cells might be considered as a pool of mesenchymal stromal cells

As an example, in the gastrointestinal tract and the urinary bladder, the subepithelial plexus formed by TCs likely supports the stem cell proliferation and epithelial renewal and, the TCs located in the muscle coat, might differentiate into ICCs and, undergoing to phenotypic changes, become a cell source of gastrointestinal stromal tumors such as GIST. A shift toward a myofibroblast phenotype has been proposed for the TCs located in the urinary bladder lamina propria.

In the female genital tract, although there are no reported interactions with stem cells, a role for TCs in this direction cannot be overlooked because neo-angiogenesis undoubtedly accompanies myometrial hypertrophy.

Moreover, in the heart, a special inter-relation exists between TCs and cardiac stem cells based on the exchange of information via extracellular vesicles which shuttle miRNA or by direct connections through typical and atypical junctions. The secretome of TCs might enhance the proliferation and differentiation of cardiac stem cells. Suggestions were also made regarding the possibility of their role in the re-activation of dormant myocardial precursors during the repair of the adult heart, while in embryo TCs act as inductors/regulators of differentiation during morphogenesis.

3. The TCs scaffold located in all the hollow organs follows organ distension and relaxation likely to avoid anomalous organ deformation and to control blood vessels closure or rheology. This is *a mechanical role* whose importance has been demonstrated in some gastrointestinal pathologies, where TCs loss provokes severe architectural derangement and contributes to the altered 3D ECM organization in the fibrotic intestinal wall. In the uterus, TCs can function as a sensor for the mechanical stress exerted on the uterine wall, allowing uniform uterine growth during pregnancy, by mechanosensitive coordination due to the existence of different ionic channels which can be modulated by pharmacological interventions.

Moreover, TCs can also be regarded as chemical sensors as it was hypothesized for the human uterus and fallopian tube, where TCs might play an important role in the uterine contraction mechanism due to the presence of estrogen and progesterone receptors at their level.

4. Because of the anatomical complexity of the hollow organs and the great variety of cell populations interacting with the TCs, many different and organ-specific TCs roles are conceivable suggesting that these cells might take center stage in *the integration of the overall interstitial information* from the vascular, nervous, and immune systems, as well as from tissue-resident stem cells

In the gastrointestinal tract, a particular role is played by the intramuscular hetero-cellular TCs network in *supporting the spreading of the slow waves generated by the ICCs*, which are electrically coupled to the smooth muscle cells, thus contributing to the regulation of gastrointestinal motility. In agreement with this hypothesis is the evidence that the simultaneous reductions in the TCs and ICCs account for the intestinal dysmotility characterizing the IBD. In the urinary bladder, the sub-urothelial TCs likely play a role as intermediaries in propagating chemical or electrical stimuli locally generated being the target of the paracrine activity of the urothelium and the nervous system. The importance of these roles is plenty understood since the ULP and the urothelium constitute a sensory system capable of perceiving mechanical and chemical stimuli and whose integrated responses control the efferent pathways on the detrusor and the micturition, responses that are lost in urinary pathologies such as the NDO.

In female genital tract, TCs seems to lack in regular slow waves indicating that they are not involved in triggering or supporting the peristalsis of these organs, but more detailed studies are necessary.

Currently, for a more undoubted TCs identification and careful characterization, no single specific immunophenotype marker is adequate. Transmission electron microscope is still considered the instrument necessary for the identification of TCs with certainty. Recently, FIB-SEM tomography was used to confirm that TCs come in contact with each other thus forming networks that constitute the scaffold organizing the stromal and the parenchymal components of the organs. For in-depth studies of TCs roles, it is highly needed to find out functional markers or receptors for the specific TC subtypes and TCs functional molecules.

Conflict of interest

The authors declare no conflict of interest.

Author details

Dragos Cretoiu[1,2†], Maria Giuliana Vannucchi[3†], Yihua Bei[4†], Mirko Manetti[3], Maria Simonetta Faussone-Pellegrini[3], Lidia Ibba-Manneschi[3], Junjie Xiao[4] and Sanda Maria Cretoiu[1*]

1 Department of Cell and Molecular Biology and Histology, Carol Davila University of Medicine and Pharmacy, Bucharest, Romania

2 Alessandrescu-Rusescu National Institute of Mother and Child Health, Fetal Medicine Excellence Research Center, Bucharest, Romania

3 Department of Experimental and Clinical Medicine, Section of Anatomy and Histology, University of Florence, Florence, Italy

4 Cardiac Regeneration and Ageing Lab, Institute of Cardiovascular Sciences, School of Life Science, Shanghai University, Shanghai, China

*Address all correspondence to: sanda@cretoiu.ro

† These authors are contributed equally.

References

[1] Popescu LM et al. Interstitial cells of Cajal in pancreas. Journal of Cellular and Molecular Medicine. 2005;**9**(1):169-190

[2] Ciontea SM et al. C-kit immunopositive interstitial cells (Cajal-type) in human myometrium. Journal of Cellular and Molecular Medicine. 2005;**9**(2):407-420

[3] Popescu LM et al. Novel type of interstitial cell (Cajal-like) in human fallopian tube. Journal of Cellular and Molecular Medicine. 2005;**9**(2):479-523

[4] Cretoiu D et al. Interstitial Cajal-like cells (ICLC) as steroid hormone sensors in human myometrium: Immunocytochemical approach. Journal of Cellular and Molecular Medicine. 2006;**10**(3):789-795

[5] Popescu LM, Ciontea SM, Cretoiu D. Interstitial Cajal-like cells in human uterus and fallopian tube. Annals of the New York Academy of Sciences. 2007;**1101**:139-165

[6] Hutchings G et al. Myometrial interstitial cells and the coordination of myometrial contractility. Journal of Cellular and Molecular Medicine. 2009;**13**(10):4268-4282

[7] Popescu LM, Faussone-Pellegrini MS. Telocytes—A case of serendipity: The winding way from interstitial cells of Cajal (ICC), via interstitial Cajal-like cells (ICLC) to telocytes. Journal of Cellular and Molecular Medicine. 2010;**14**(4):729-740

[8] Vannucchi MG, Faussone-Pellegrini MS. The telocyte subtypes. Advances in Experimental Medicine and Biology. 2016;**913**:115-126

[9] Cretoiu SM. Telocytes in focus. Seminars in Cell and Developmental Biology. 2016;**55**:1-2

[10] Cretoiu SM, Cretoiu D, Popescu LM. Human myometrium - the ultrastructural 3D network of telocytes. Journal of Cellular and Molecular Medicine. 2012;**16**(11):2844-2849

[11] Popescu LM et al. Telocytes and putative stem cells in the lungs: Electron microscopy, electron tomography and laser scanning microscopy. Cell and Tissue Research. 2011;**345**(3):391-403

[12] Cretoiu D et al. Telocytes, a distinct type of cell among the stromal cells present in the lamina propria of jejunum. Histology and Histopathology. 2012;**27**(8):1067-1078

[13] Cretoiu D. The third dimension of telocytes revealed by FIB-SEM tomography. Advances in Experimental Medicine and Biology. 2016;**913**:325-334

[14] Bassotti G et al. Cellular and molecular basis of chronic constipation: Taking the functional/idiopathic label out. World Journal of Gastroenterology. 2013;**19**(26):4099-4105

[15] Albulescu R et al. The secretome of myocardial telocytes modulates the activity of cardiac stem cells. Journal of Cellular and Molecular Medicine. 2015;**19**(8):1783-1794

[16] Cretoiu D et al. Telocytes and their extracellular vesicles-evidence and hypotheses. International Journal of Molecular Sciences. 2016;**17**(8):1322. Available from: http://blog.mdpi.com/2015/12/01/a-new-look-for-mdpi-papers/

[17] Cretoiu D, Cretoiu SM. Telocytes in the reproductive organs: Current understanding and future challenges. Seminars in Cell and Developmental Biology. 2016;**55**:40-49

[18] Faussone-Pellegrini MS, Gherghiceanu M. Telocyte's contacts.

Seminars in Cell and Developmental Biology. 2016;**55**:3-8

[19] Zheng M et al. Variations of chromosomes 2 and 3 gene expression profiles among pulmonary telocytes, pneumocytes, airway cells, mesenchymal stem cells and lymphocytes. Journal of Cellular and Molecular Medicine. 2014;**18**(10):2044-2060

[20] Zheng Y et al. Protein profiling of human lung telocytes and microvascular endothelial cells using iTRAQ quantitative proteomics. Journal of Cellular and Molecular Medicine. 2014;**18**(6):1035-1059

[21] Sun X et al. Differences in the expression of chromosome 1 genes between lung telocytes and other cells: Mesenchymal stem cells, fibroblasts, alveolar type II cells, airway epithelial cells and lymphocytes. Journal of Cellular and Molecular Medicine. 2014;**18**(5):801-810

[22] Song D et al. Comparison of chromosome 4 gene expression profile between lung telocytes and other local cell types. Journal of Cellular and Molecular Medicine. 2016;**20**(1):71-80

[23] Shoshkes-Carmel M et al. Subepithelial telocytes are an important source of Wnts that supports intestinal crypts. Nature. 2018;**557**(7704):242-246

[24] Rosa I et al. Morphological evidence of telocytes in human synovium. Scientific Reports. 2018;**8**(1):3581

[25] Marini M et al. Reappraising the microscopic anatomy of human testis: Identification of telocyte networks in the peritubular and intertubular stromal space. Scientific Reports. 2018;**8**(1):14780

[26] Varga I et al. Recently discovered interstitial cells "telocytes" as players in the pathogenesis of uterine leiomyomas. Medical Hypotheses. 2018;**110**:64-67

[27] Varga I et al. Recently discovered interstitial cell population of telocytes: Distinguishing facts from fiction regarding their role in the pathogenesis of diverse diseases called "telocytopathies". Medicina (Kaunas, Lithuania). 2019;**55**(2):56

[28] Cretoiu SM. Immunohistochemistry of telocytes in the uterus and fallopian tubes. Advances in Experimental Medicine and Biology. 2016;**913**:335-357

[29] Diaz-Flores L et al. CD34+ stromal cells/fibroblasts/fibrocytes/telocytes as a tissue reserve and a principal source of mesenchymal cells. Location, morphology, function and role in pathology. Histology and Histopathology. 2014;**29**(7):831-870

[30] Marini M et al. Telocytes in skeletal, cardiac and smooth muscle interstitium: Morphological and functional aspects. Histology and Histopathology. 2018;**33**(11):1151-1165

[31] Cretoiu SM et al. Interstitial Cajal-like cells of human fallopian tube express estrogen and progesterone receptors. Journal of Molecular Histology. 2009;**40**(5-6):387-394

[32] Vannucchi MG et al. Telocytes express PDGFRalpha in the human gastrointestinal tract. Journal of Cellular and Molecular Medicine. 2013;**17**(9):1099-1108

[33] Vannucchi MG et al. Telocytes subtypes in human urinary bladder. Journal of Cellular and Molecular Medicine. 2014;**18**(10):2000-2008

[34] Vannucchi MG, Traini C. Interstitial cells of Cajal and telocytes in the gut: Twins, related or simply neighbor cells? Biomolecular Concepts. 2016;**7**(2):93-102

[35] Vannucchi MG, Traini C. The telocytes/myofibroblasts 3-D network forms a stretch receptor in the human

bladder mucosa. Is this structure involved in the detrusor overactive diseases? Annals of Anatomy. 2018;**218**:118-123

[36] Popescu LM et al. The connective connection: Interstitial cells of Cajal (ICC) and ICC-like cells establish synapses with immunoreactive cells. Electron microscope study in situ. Journal of Cellular and Molecular Medicine. 2005;**9**(3):714-730

[37] Vannucchi MG, Bani D, Faussone-Pellegrini MS. Telocytes contribute as cell progenitors and differentiation inductors in tissue regeneration. Current Stem Cell Research and Therapy. 2016;**11**(5):383-389

[38] Traini C et al. Adaptive changes of telocytes in the urinary bladder of patients affected by neurogenic detrusor overactivity. Journal of Cellular and Molecular Medicine. 2018;**22**(1):195-206

[39] Kurahashi M et al. A novel population of subepithelial platelet-derived growth factor receptor alpha-positive cells in the mouse and human colon. American Journal of Physiology. Gastrointestinal and Liver Physiology. 2013;**304**(9):G823-G834

[40] Faussone-Pellegrini MS, Popescu LM. Telocytes. Biomolecular Concepts. 2011;**2**(6):481-489

[41] Ibba-Manneschi L, Rosa I, Manetti M. Telocyte implications in human pathology: An overview. Seminars in Cell and Developmental Biology. 2016;**55**:62-69

[42] Sanders KM, Ward SM, Koh SD. Interstitial cells: Regulators of smooth muscle function. Physiological Reviews. 2014;**94**(3):859-907

[43] Gibbons SJ et al. Apoptotic cell death of human interstitial cells of Cajal. Neurogastroenterology and Motility. 2009;**21**(1):85-93

[44] Huizinga JD, White EJ. Progenitor cells of interstitial cells of Cajal: On the road to tissue repair. Gastroenterology. 2008;**134**(4):1252-1254

[45] Rieder F, Fiocchi C. Intestinal fibrosis in inflammatory bowel disease—Current knowledge and future perspectives. Journal of Crohn's and Colitis. 2008;**2**(4):279-290

[46] Ibba-Manneschi L, Rosa I, Manetti M. Telocytes in chronic inflammatory and fibrotic diseases. Advances in Experimental Medicine and Biology. 2016;**913**:51-76

[47] Latella G et al. Can we prevent, reduce or reverse intestinal fibrosis in IBD? European Review for Medical and Pharmacological Sciences. 2013;**17**(10):1283-1304

[48] Wang XY et al. Ultrastructural injury to interstitial cells of Cajal and communication with mast cells in Crohn's disease. Neurogastroenterology and Motility. 2007;**19**(5):349-364

[49] Bernardini N et al. Immunohistochemical analysis of myenteric ganglia and interstitial cells of Cajal in ulcerative colitis. Journal of Cellular and Molecular Medicine. 2012;**16**(2):318-327

[50] Milia AF et al. Telocytes in Crohn's disease. Journal of Cellular and Molecular Medicine. 2013;**17**(12):1525-1536

[51] Manetti M et al. Telocytes are reduced during fibrotic remodelling of the colonic wall in ulcerative colitis. Journal of Cellular and Molecular Medicine. 2015;**19**(1):62-73

[52] Richter M, Kostin S. The failing human heart is characterized by decreased numbers of telocytes as result of apoptosis and altered extracellular

matrix composition. Journal of Cellular and Molecular Medicine. 2015;**19**(11):2597-2606

[53] Manetti M et al. Evidence for progressive reduction and loss of telocytes in the dermal cellular network of systemic sclerosis. Journal of Cellular and Molecular Medicine. 2013;**17**(4):482-496

[54] Manetti M et al. A loss of telocytes accompanies fibrosis of multiple organs in systemic sclerosis. Journal of Cellular and Molecular Medicine. 2014;**18**(2):253-262

[55] Ricci R et al. Telocytes are the physiological counterpart of inflammatory fibroid polyps and PDGFRA-mutant GISTs. Journal of Cellular and Molecular Medicine. 2018;**22**(10):4856-4862

[56] Cretoiu SM, Popescu LM. Telocytes revisited. Biomolecular Concepts. 2014;**5**(5):353-369

[57] Fry CH, Vahabi B. The role of the mucosa in normal and abnormal bladder function. Basic and Clinical Pharmacology and Toxicology. 2016;**119**(Suppl 3):57-62

[58] Abrams P et al. The standardisation of terminology in lower urinary tract function: Report from the standardisation sub-committee of the international continence society. Urology. 2003;**61**(1):37-49

[59] Wiseman OJ, Fowler CJ, Landon DN. The role of the human bladder lamina propria myofibroblast. BJU International. 2003;**91**(1):89-93

[60] Gevaert T et al. Morphometric and quantitative immunohistochemical analysis of disease-related changes in the upper (suburothelial) lamina propria of the human bladder dome. PLoS One. 2015;**10**(5):e0127020

[61] Roosen A et al. Suburothelial myofibroblasts in the human overactive bladder and the effect of botulinum neurotoxin type A treatment. European Urology. 2009;**55**(6):1440-1448

[62] Hatta K et al. Culture of rat endometrial telocytes. Journal of Cellular and Molecular Medicine. 2012;**16**(7):1392-1396

[63] Aleksandrovych V, Walocha JA, Gil K. Telocytes in female reproductive system (human and animal). Journal of Cellular and Molecular Medicine. 2016;**20**(6):994-1000

[64] Shafik A et al. Identification of a vaginal pacemaker: An immunohistochemical and morphometric study. Journal of Obstetrics and Gynaecology. 2007;**27**(5):485-488

[65] Ullah S et al. Identification and characterization of telocytes in the uterus of the oviduct in the Chinese soft-shelled turtle, Pelodiscus sinensis: TEM evidence. Journal of Cellular and Molecular Medicine. 2014;**18**(12):2385-2392

[66] Cretoiu SM et al. Complex effects of imatinib on spontaneous and oxytocin-induced contractions in human non-pregnant myometrium. Acta Physiologica Hungarica. 2011;**98**(3):329-338

[67] Cretoiu SM et al. Isolated human uterine telocytes: Immunocytochemistry and electrophysiology of T-type calcium channels. Histochemistry and Cell Biology. 2015;**143**(1):83-94

[68] Cantarero I et al. Transmission electron microscopy as key technique for the characterization of telocytes. Current Stem Cell Research and Therapy. 2016;**11**(5):410-414

[69] Janas P et al. Telocytes in the female reproductive system: An overview of

up-to-date knowledge. Advances in Clinical and Experimental Medicine. 2018;**27**(4):559-565

[70] Cretoiu SM et al. Telocytes: Ultrastructural, immunohistochemical and electrophysiological characteristics in human myometrium. Reproduction. 2013;**145**(4):357-370

[71] Roatesi I et al. Uterine telocytes: A review of current knowledge. Biology of Reproduction. 2015;**93**(1):10

[72] Mirancea N. Telocyte—A particular cell phenotype. Infrastructure, relationships and putative functions. Romanian Journal of Morphology and Embryology. 2016;**57**(1):7-21

[73] Rusu MC et al. Telocytes of the human adult trigeminal ganglion. Cell Biology and Toxicology. 2016;**32**(3):199-207

[74] Kostin S. Cardiac telocytes in normal and diseased hearts. Seminars in Cell and Developmental Biology. 2016;**55**:22-30

[75] Cretoiu D et al. Telocytes heterogeneity: From cellular morphology to functional evidence. Seminars in Cell and Developmental Biology. 2017;**64**:26-39

[76] Campeanu RA et al. Near-infrared low-level laser stimulation of telocytes from human myometrium. Lasers in Medical Science. 2014;**29**(6):1867-1874

[77] Rosenbaum ST et al. Immunolocalization and expression of small-conductance calcium-activated potassium channels in human myometrium. Journal of Cellular and Molecular Medicine. 2012;**16**(12):3001-3008

[78] Banciu A et al. Beta-estradiol regulates voltage-gated calcium channels and estrogen receptors in telocytes from human myometrium. International Journal of Molecular Sciences. 2018;**19**(5):1413

[79] Radu BM et al. Calcium signaling in interstitial cells: Focus on telocytes. International Journal of Molecular Sciences. 2017;**18**(2):397

[80] Diaz-Flores L et al. Behaviour of telocytes during physiopathological activation. Seminars in Cell and Developmental Biology. 2016;**55**:50-61

[81] Xiao J et al. Telocytes in exercise-induced cardiac growth. Journal of Cellular and Molecular Medicine. 2016;**20**(5):973-979

[82] Zheng Y et al. Human lung telocytes could promote the proliferation and angiogenesis of human pulmonary microvascular endothelial cells in vitro. Molecular and Cellular Therapies. 2014;**2**:3

[83] Ma R et al. Telocytes promote VEGF expression and alleviate ventilator-induced lung injury in mice. Acta Biochimica et Biophysica Sinica. 2018;**50**(8):817-825

[84] Enciu AM, Popescu LM. Telopodes of telocytes are influenced in vitro by redox conditions and ageing. Molecular and Cellular Biochemistry. 2015;**410**(1-2):165-174

[85] Jiang XJ et al. An in vitro investigation of telocytes-educated macrophages: Morphology, heterocellular junctions, apoptosis and invasion analysis. Journal of Translational Medicine. 2018;**16**(1):85

[86] Yang P et al. Features of telocytes in agricultural animals. Advances in Experimental Medicine and Biology. 2016;**913**:105-113

[87] Yang P et al. Cellular evidence of telocytes as novel interstitial cells within the magnum of chicken oviduct. Cell Transplantation. 2017;**26**(1):135-143

[88] Yang XJ et al. Telocytes damage in endometriosis-affected rat oviduct and potential impact on fertility. Journal

of Cellular and Molecular Medicine. 2015;**19**(2):452-462

[89] Yang J et al. Ultrastructure damage of oviduct telocytes in rat model of acute salpingitis. Journal of Cellular and Molecular Medicine. 2015;**19**(7):1720-1728

[90] Varga I et al. Functional histology and possible clinical significance of recently discovered telocytes inside the female reproductive system. Archives of Gynecology and Obstetrics. 2016;**294**(2):417-422

[91] Urban L et al. Telocytes (interstitial Cajal-like cells) in human fallopian tubes. Bratislavské Lekárske Listy. 2016;**117**(5):263-267

[92] Aleksandrovych V et al. The autonomic innervation and uterine telocyte interplay in leiomyoma formation. Cell Transplantation. 2019;**28**(5):619-629

[93] Benagiano G, Brosens I. The endometrium in adenomyosis. Womens Health (London). 2012;**8**(3): 301-312

[94] Brosens I, Benagiano G. Endometriosis, a modern syndrome. The Indian Journal of Medical Research. 2011;**133**:581-593

[95] Horne AW et al. Placental growth factor: A promising diagnostic biomarker for tubal ectopic pregnancy. The Journal of Clinical Endocrinology and Metabolism. 2011;**96**(1):E104-E108

[96] Bei Y et al. Telocytes in regenerative medicine. Journal of Cellular and Molecular Medicine. 2015;**19**(7):1441-1454

[97] Horch RE et al. The potential role of telocytes for tissue engineering and regenerative medicine. Advances in Experimental Medicine and Biology. 2016;**913**:139-147

[98] Dixon RE et al. Time-dependent disruption of oviduct pacemaker cells by chlamydia infection in mice. Biology of Reproduction. 2010;**83**(2):244-253

[99] Yang XJ. Telocytes in inflammatory gynaecologic diseases and infertility. Advances in Experimental Medicine and Biology. 2016;**913**:263-285

[100] Abd-Elhafeez HH, Soliman SA. New description of telocyte sheaths in the bovine uterine tube: An immunohistochemical and scanning microscopic study. Cells, Tissues, Organs. 2017;**203**(5):295-315

[101] Bei Y et al. Telocytes in cardiac regeneration and repair. Seminars in Cell and Developmental Biology. 2016;**55**:14-21

[102] Yang Y et al. Telocytes in human heart valves. Journal of Cellular and Molecular Medicine. 2014;**18**(5):759-765

[103] Tay H, Vandecasteele T, Van den Broeck W. Identification of telocytes in the porcine heart. Anatomia, Histologia, Embryologia. 2017;**46**(6):519-527

[104] Cretoiu D et al. Human cardiac telocytes: 3D imaging by FIB-SEM tomography. Journal of Cellular and Molecular Medicine. 2014;**18**(11):2157-2164

[105] Sheng J et al. Electrophysiology of human cardiac atrial and ventricular telocytes. Journal of Cellular and Molecular Medicine. 2014;**18**(2):355-362

[106] Bei Y et al. Cardiac telocytes and fibroblasts in primary culture: Different morphologies and immunophenotypes. PLoS One. 2015;**10**(2):e0115991

[107] Zhou Q et al. Cardiac telocytes are double positive for CD34/PDGFR-alpha. Journal of Cellular and Molecular Medicine. 2015;**19**(8):2036-2042

[108] Li YY et al. Comparative analysis of telomerase activity in CD117(+) CD34(+) cardiac telocytes with bone mesenchymal stem cells, cardiac fibroblasts and cardiomyocytes. Chinese Medical Journal. 2015;**128**(14):1942-1947

[109] Kostin S. Myocardial telocytes: A specific new cellular entity. Journal of Cellular and Molecular Medicine. 2010;**14**(7):1917-1921

[110] Faussone-Pellegrini MS, Bani D. Relationships between telocytes and cardiomyocytes during pre- and post-natal life. Journal of Cellular and Molecular Medicine. 2010;**14**(5):1061-1063

[111] Shi J et al. miR-17-3p contributes to exercise-induced cardiac growth and protects against myocardial ischemia-reperfusion injury. Theranostics. 2017;7(3):664-676

[112] Liu X et al. miR-222 is necessary for exercise-induced cardiac growth and protects against pathological cardiac remodeling. Cell Metabolism. 2015;**21**(4):584-595

[113] Mandache E et al. Telocytes in human isolated atrial amyloidosis: Ultrastructural remodelling. Journal of Cellular and Molecular Medicine. 2010;**14**(12):2739-2747

[114] Nour MS et al. Histological and immunohistochemical study of cardiac telocytes in a rat model of isoproterenol-induced myocardial infarction with a reference to the effect of grape seed extract. Acta Histochemica. 2017;**119**(7):747-758

[115] Zhao B et al. Cardiac telocytes were decreased during myocardial infarction and their therapeutic effects for ischaemic heart in rat. Journal of Cellular and Molecular Medicine. 2013;**17**(1):123-133

[116] Zhao B et al. Intramyocardial transplantation of cardiac telocytes decreases myocardial infarction and improves post-infarcted cardiac function in rats. Journal of Cellular and Molecular Medicine. 2014;**18**(5):780-789

Section 4

Stem Cells

Chapter 5

Innovations in Human Stem Cell Research: A Holy Grail for Regenerative Medicine

Yanling Liao, Hongwen Zhu, Larisa Ivanova and Mitchell S. Cairo

Abstract

Stem cells are unspecialized cells capable of renewing themselves and giving rise to differentiated and specialized cell subtypes. There are two general categories of stem cells, i.e., pluripotent stem cells capable of differentiation into any cell type in the human body and multipotent adult stem cells maintaining tissue homeostasis in postnatal life. Investigations in both these categories of stem cells have expanded our knowledge on human organogenesis and tissue regeneration and have suggested potential therapeutic functions of stem cells in regenerative medicine. The advent of induced pluripotent stem cell (iPSC) technology a decade ago further revolutionized stem cell biology and has given rise to the translation of stem cell-based therapies. This chapter will summarize some of the exciting progress and challenges in the applications of iPSC-derived stem cells and adult stem cells and the potential of translational and clinical research of these stem cells in regenerative medicine.

Keywords: stem cells, regenerative therapy, clinical studies, iPSCs, cord blood, adult stem cells

1. Introduction

Stem cells are a unique subset of cells that possess the biological properties of self-renewal and differentiation. The term "Stammzelle" (German for stem cell) was first used by a German biologist Ernst Haeckel in 1868 regarding unicellular organisms as the phylogenetic ancestors of multicellular organisms [1]. In 1877, Haeckel further applied the notion of stem cells to describe the fertilized egg cell as the cell of origin for all the cells in an organism [2].

Indeed, there are stem cells of different potency that not only form the foundation of any organisms but also throughout the life of the organism maintain tissue homeostasis. Following egg fertilization with a sperm, a zygote is generated (**Figure 1A**). This single-cell zygote and the cells that immediately arise in the first few divisions are totipotent stem cells, as they are capable of becoming a whole embryo including the extra-embryonic tissue, i.e., placenta. After about 5 days of human embryonic development, the zygote develops into a blastocyst. The preimplantation blastocyst consists of a mostly hollow ball of cells, with the outer cell layer as trophoblast that develops into the placenta and the inner cell mass (ICM) that is pluripotent and gives rise to all cell types in the body. These pluripotent stem

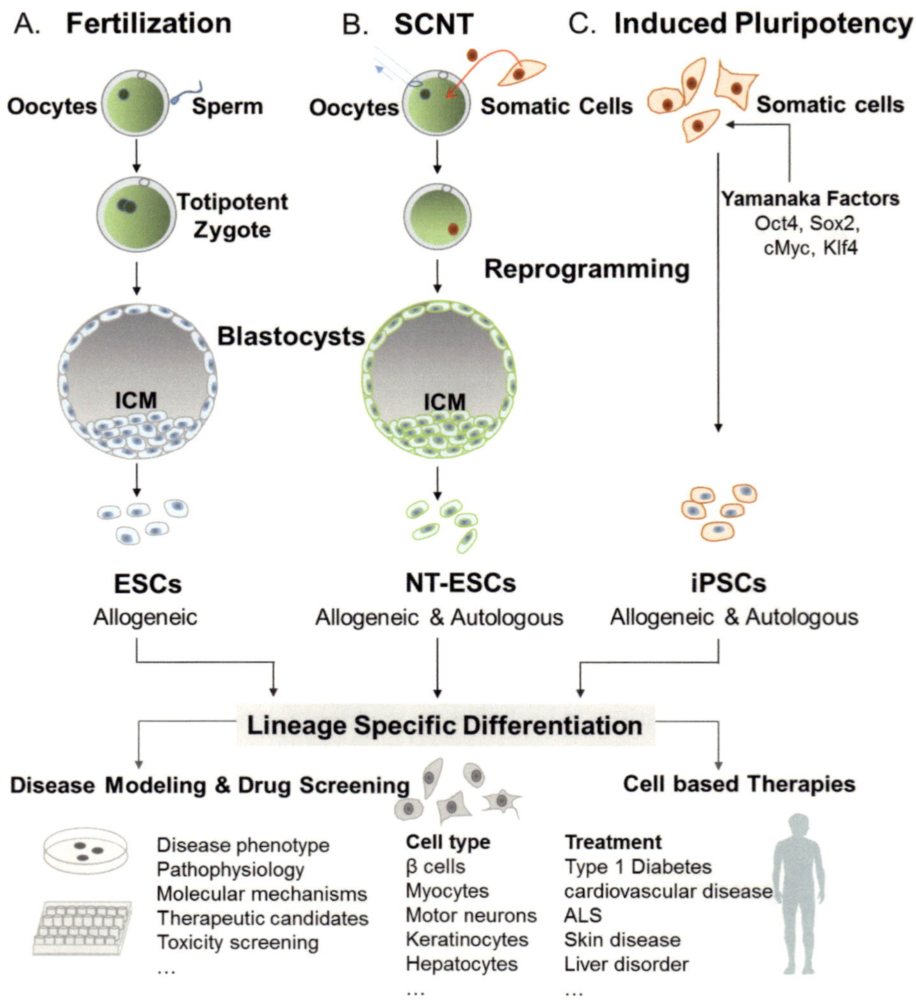

Figure 1.
Derivation of pluripotent stem cells (PSCs) including embryonic stem cells (ESCs) from fertilization of oocytes and sperm (A), nuclear transfer ESCs (NT-ESCs) from somatic cell nuclear transfer (SCNT) of somatic cell nuclei into denucleated oocytes (B), and induced pluripotent stem cells from forced expression of Yamanaka transcription factors in somatic cells (C). All the three types of PSCs can be propagated extensively in vitro and undergo directed differentiation into any cell type of the body, which can be utilized in disease modeling and drug screening and developed as cell-based therapies. As generation of ESCs involve disruption of the embryos, they can only be used as an allogenic source, while NT-ESCs and iPSCs can be developed in both allogenic and autologous settings.

cells (PSCs) in the ICM are the source for embryonic stem (ES) cells [3]. Apart from PSCs, there are tissue-specific stem cells residing in several organs such as the skin, gut, blood, and brain, also referred to as "adult" or "somatic" stem cells with self-renewal and multipotent differentiation capacity that are responsible for most regenerative activities throughout the life of the organism.

2. Pluripotent stem cells (PSCs)

2.1 ES cells

The culture of ES cells from mouse blastocysts was first reported by Evans et al. in 1981 and herein opened a new era of research in gene targeting to model human diseases [3]. Based on this technology, Martin Evans, Mario Capecchi, and Oliver Smithies shared the 2007 Nobel Prize for their discoveries of principles for introducing specific gene modifications in mice by the use of ES cells. In 1998, Thomson and collaborators reported the derivation of ES cell lines from human blastocysts produced by in vitro fertilization (IVF) [4]. As these human ES cells are immortal and can give rise to all cell types in the body, they are invaluable tools to dissect the molecular and genetic events in human development and also allow drug testing in a wide range of diseases (**Figure 1**). Moreover, human ES cells provide an unlimited source to derive cells and tissues that could be used as cell-based therapies to treat degenerative diseases such as heart disease and type 1 diabetes (**Figure 1**). Indeed, detailed differentiation protocols have been developed to derive specific mature cells from human ES cells, for example, dopaminergic neurons, cardiomyocytes, β cells, and keratinocytes as therapies for Parkinson's disease, myocardial infarction, type I diabetes, and skin disease, respectively. However, as the derivation of ES cells involves destruction of human embryos, even though the embryos were produced by IVF and would otherwise be discarded, it nevertheless has raised ethical concerns and elicited controversial debates on the use of human ES cells [5]. Another hurdle for the therapeutic use of ES-derived cells is that they can only be used as allogeneic cells; thus they are subject to immune-based rejection in their recipients [6].

2.2 PSCs derived from reprogramming

The 2012 Nobel Prize in Physiology or Medicine was awarded to two scientists, Sir John B. Gurdon and Shinya Yamanaka, PhD, whose stem cell research was separated by over four decades but both demonstrated that the developmental clock of a specialized cell can be turned back to an immature (pluripotent) state. In John B. Gurdon's classic experiment in 1962, he replaced the nucleus in an egg cell of a frog with the nucleus from a mature intestinal cell [7]. The modified egg cell subsequently developed into a normal tadpole. Using the same principle of Gurdon's work which is known as somatic cell nuclear transfer (SCNT), the first mammalian clone "Dolly the Sheep" was born in 1996, following transfer of the nucleus from a mammary gland into an enucleated sheep egg [8]. Therefore, factors presented in mature, metaphase II-arrested oocyte cytoplasm are capable of reprogramming somatic cell nuclei to an undifferentiated state. This unique ability of oocytes has led to an important application of utilizing SCNT in therapeutic cloning to generate ES cells (ESCs) (NT-ESCs) for regenerative medicine (**Figure 1B**).

Forty-four years after Gurdon's discovery, Shinya Yamanaka took a different approach and transduced the differentiated mammalian cells, i.e., fibroblasts with a cocktail of reprogramming factors (Oct3/4, Sox2, Klf4, and c-Myc) using retroviruses (**Figure 1C**) [9]. About 30 days after transduction, induced pluripotent stem cells (iPSCs) with properties similar to ES cells were generated at an efficiency of ~0.02% [9]. After the initial discovery of Yamanaka's reprogramming techniques, several other reprogramming methods, such as transduction with nonintegrating viruses (adenovirus and Sendai virus) or transfection with protein, mRNA, Piggy Bac element, minicircle vectors, and episomal plasmids, have been developed to produce iPSCs without integration of exogenous genes to the genome of recipient cells (reviewed in [10]). iPSCs can also be generated from more easily accessible cell sources such as blood [11].

Therefore, the PSCs produced by SCNT (NT-ESCs) and forced expression of reprogramming transcription factors (iPSCs) are both alternatives to ESCs in regenerative medicine. Epigenetic and transcriptomic comparisons between isogenic iPSCs and NT-ESCs have demonstrated that NT-ESCs more closely resemble bona fide ESCs derived from fertilized embryos [12, 13]. Moreover, a most important difference between iPSCs and NT-ESCs is the source of mitochondrial DNA (mtDNA). The mtDNA in NT-ESCs is of an oocyte germline origin, while in iPSCs is of a parental somatic origin. Due to the random nature of somatic mtDNA mutations, the frequency of mtDNA defects in iPSCs has been demonstrated to increase with the age of somatic cells [14]. Thus, NT-ESCs, carrying mutation-free mtDNA and closely resembling ESCs, represent an invaluable stem cell source for regenerative medicine. However, as the derivation of NT-ESCs requires donor oocytes, which are more technically challenging than iPSCs and are also subject to ethical and/or legal restrictions, the majority of current PSC research has been carried out using iPSCs.

2.3 iPSC disease modeling

These groundbreaking discoveries have revolutionized our understanding of stem cell development and created novel opportunities for human disease modeling and drug screening in "disease-in-a-dish" models (**Figure 1**) [15]. To date, significant progress has been made utilizing human iPSCs to model various neurological disorders, inherited heart diseases, and other genetic diseases such as Duchene muscular dystrophy and recessive dystrophic epidermolysis bullosa (RDEB) [16, 17]. Utilizing amyotrophic lateral sclerosis (ALS), also called Lou Gehrig's disease as an example, ALS is a neurodegenerative disease that primarily affects corticospinal "upper" motor neurons (UMNs) and spinal cord "lower" motor neurons (LMNs), resulting in progressive muscle weakness [18]. In about 10% of patients with ALS, the disease runs in the family (familiar ALS) with mutations in around 20 genes including SOD1, TARDBP, FIS, and C9orf72 identified as common causes [18]. The remaining 90% of the patients are classified as sporadic ALS, with the causative mutations largely unidentified. As the iPSCs generated from ALS patients and differentiated into motor neurons carry the same genetic background as the patients, it represents a novel tool for studying disease pathology of ALS, particularly the sporadic form, which is not possible in the other model systems. A proof-of-principle study on derivation of iPSCs from an ALS patient and differentiation into LMNs was reported in 2008 [19]. Subsequently, Kiskinis et al. and Chen et al., respectively, established in vitro models of ALS by generating iPSC-derived LMNs from patients carrying different SOD1 mutations. Both studies recapitulated the spontaneous and progressive decrease in cell viability and ALS-related morphological changes including reduction in soma size and altered dendrites, which was linked to neurofilament aggregation [20, 21]. Chen et al. further demonstrated the pathological features of mutated SOD1 in patient-derived MNs, but not in non-MNs. Only in MNs, mutated SOD1 bound to the 3′UTR region of neurofilament (NF)-L mRNA resulted in neurofilament aggregation, restoring the expression of NF-L mitigated neurite degeneration of the ALS-iPSC-derived MNs. Meanwhile, Wainger et al. generated iPSC-derived LMNs from patients carrying SOD1, C9orf72, or FUS mutations [22]. All these ALS-iPSC-derived LMNs with distinct genetic mutations have recapitulated essential disease features and discovered common molecular pathways driving ALS pathogenesis, opening the possibility of new and effective drug screening [23]. However, challenges still remain for in vitro modeling for ALS using iPSCs [18]. Different protocols have been reported in deriving LMNs from iPSCs; thus, criteria need to be established to compare the MNs generated using different methods. Moreover, generation of UMNs from PSCs involves a series of steps and is more challenging than that of LMNs [18]. The current protocols for deriving UMNs mostly resulted in heterogeneous, neocortical-like neurons that are immature and "stalled" at a stage resembling mid-embryonic differentiation in vivo [24]. Therefore, promoting subtype-specific differentiation and maturation will be crucial to an accurate ALS modeling. Indeed, the abilities of iPSC-derived cells to exhibit maturation and aging are crucial for accurate in vitro modeling of all the adult-onset diseases.

2.4 iPSC-based therapies

The most significant advantage of iPSCs lies in its application in cell-based therapies. iPSCs can be developed without destroying human embryos, therefore circumventing the ethical obstacles of utilizing and generating human ES cells. Being able to differentiate into all cell types in the body similar to ES cells, iPSCs theoretically provide an unlimited source of cells for autologous transplantation, eliminating the need for immunosuppression. Moreover, scientists have established robust directed differentiation protocols with sequential activation and inhibition of molecular differentiation pathways to generate a wide range of somatic cells from iPSCs, such as β cells and cardiomyocytes (**Figure 1**).

A challenge for the PSC or iPSC differentiation, as also mentioned above in the iPSC-derived MNs, is that the PSC-derived cells tend to be immature. This is indeed the major limitation for translating iPSC-derived red blood cells into the clinic [25]. In 2008, Lu et al. reported differentiation of human ES cells into functional oxygen-carrying erythrocytes on a large scale with up to 60% enucleation rate [26]. In comparison, differentiation of iPSCs along the erythroid lineage generated orthochromatic (nucleated) erythroblasts and reticulocytes. In most reports, the differentiated red blood cells express mainly fetal and embryonic globins, but very little adult-type (β-) globin [27]. This is likely due to the low level of erythroid Kruppel-like factor 1 (EKLF1) and absence of BCL11A in these iPSC-derived red blood cells. These two factors have been demonstrated to be essential for the developmental switch from fetal to adult globin expression [28]. Inducible expression of KLF1 during later stages of the differentiation process has been recently demonstrated to enhance differentiation and maturation of red blood cells from both human ES cells and iPSCs [29].

Recent advances in the development of programmable site-specific nucleases, including zinc-finger nuclease (ZFN), transcription activator-like effector nucleases (TALENs), and the clustered regularly interspaced short palindromic repeat (CRISPR)-associated protein (Cas)9 system, have enabled target-specific introduction of transgene or correction of disease-specific mutations by homologous recombination, creating novel opportunities not only for disease modeling and drug testing but also generation of genetically corrected cells for autologous transplantation (**Figure 2**).

The development of iPSC technology has also revolutionized the future treatment for end-stage organ failure. Takebe et al. recently reported vascularized and functional mini-livers or liver buds created in vitro based on human iPSCs [30]. In this proof-of-concept demonstration, the authors first prepared hepatic endoderm cells from human iPSCs by directed differentiation. About 80% of the differentiated cells express liver-specific marker HNF4A. To recapitulate early organogenesis, the investigators next cultured the iPSC-derived hepatic endoderm cells with two stromal cell populations, i.e., human umbilical vein endothelial cells and human mesenchymal stem cells (MSCs), in a traditional two-dimensional culture condition. Intriguingly, the iPSC-derived hepatic cells

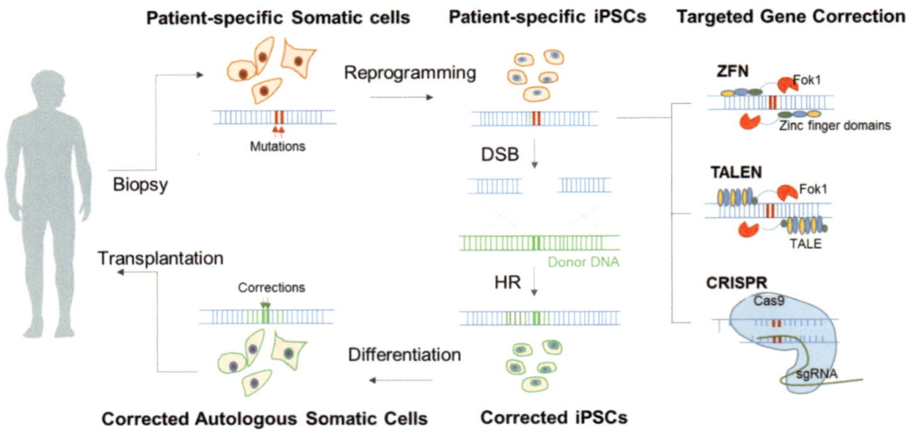

Figure 2.
Development of iPSC gene correction and autologous transplantation therapy. Biopsies such as the skin or blood can be obtained from a patient with genetic mutations and reprogrammed into patient-specific iPSCs. Through targeted nuclease technologies, such as zinc-finger nuclease which contains sequence-specific DNA-binding domain fused to a non-specific Fok1 *endonuclease enzyme, transcription activator-like effector proteins consisted of tandem DNA-binding repeats linked with the* Fok1 *enzyme, and the most recent method of CRISPR/Cas system that utilizes a single-guide RNA (sgRNA) and a protospacer adjacent motif for efficient genome targeting and binding followed by activity of the Cas enzyme, a double-stranded break (DSB) occurs at the target site of genome, inducing activation of internal DNA repair mechanism. Homologous recombination (HR) can then be achieved to incorporate exogenously transduced donor DNA to repair the mutations in the genome. After validation of the target-specific gene correction, the corrected iPSCs can be further differentiated into a cell type of preference for autologous transplantation.*

self-organized into three-dimensional cell clusters that resemble in vivo liver buds during embryonic development. Moreover, within 48 hours of transplantation into the nonobese diabetic/severe combined immunodeficiency disease (NOD/SCID) mice, the vasculatures in the iPSC-derived liver buds became functional by connecting to the host vessels, which further stimulated the maturation of iPSC-derived liver buds into the tissue resembling the adult liver. Considering critical shortage of donor organs, development of iPSC-derived organoids suggests an alternative and innovative regenerative approach for patients with end-stage organ failure.

New advances have also been made in cardiac tissue engineering for cardiovascular diseases. The protocols for differentiating ES cells or iPSCs into cardiomyocytes, smooth muscle cells, and endothelial cells that are the main functional cell types in the heart have been reported [31, 32]. However, although readily obtainable, the morphology, calcium handling, electric coupling, contraction stress, and electrophysiology of the PSC-derived cardiomyocytes have been demonstrated to be immature compared to adult cardiomyocytes (reviewed in [33]). With that, significant bioengineering efforts have been made to recapitulate environmental cues to enable maturation of newly differentiated cardiomyocytes and to promote vascular network formation (reviewed in [34]). To address the need for tissue and/or organ transplantation, there are also exciting advances in incorporating biocompatible materials, cells, and supporting components into complex 3D functional living tissues [35]. Excitingly, Noor et al. recently reported a 3D printing of thick, vascularized, and perfusable cardiac patches that fully match the immunological, biochemical, and anatomical properties of the patient [36]. This was the first report on the use of fully personalized, non-supplemented materials as bioink for 3D printing. In this study, fatty tissue was extracted from a patient. The cells from the tissue were reprogrammed to iPSCs, followed by directed differentiation into cardiomyocytes and endothelial cells. The remaining fatty tissue was decellularized and processed to generate a thermo-responsive hydrogel. The iPSC-derived cells were then encapsulated with the hydrogel and served as the bioink for 3D printing of vascularized patches and complex cellularized structures. The investigators of this study also pointed out the obstacles that need to be overcome for a more applicable 3D printing, including efficient generation of a sufficient number of cells for the organ printing, identifying biochemical and physical cues for cell maturation and conditions for long-term cultivation and a higher-resolution imaging of the entire blood vessels for the blueprint of 3D printing, etc. Nevertheless, the results from these studies have shed light on developing autologous engineered tissue or organs for transplantation.

2.5 Xenogeneic generation of human organs using PSCs

With the growing knowledge of organogenesis and the aid of gene editing technologies, scientists are also pushing the boundary and creating interspecies chimeras to grow human organs in animals, which ideally could subsequently be transplanted into people. In 2017, Yamaguchi et al. demonstrated that injection of mouse PSCs in the blastocysts of apancreatic $Pdx1^{mu/mu}$ rats (TALEN-mediated disruption in *Pdx1* gene, a master regulator for pancreas development) which resulted in generation of a mouse pancreas in the rat [37]. Moreover, when the mouse pancreas grown in the rat was transplanted into diabetic mice, they were able to cure diabetes in the recipients without the administration of immunosuppression [37]. Wu et al. subsequently reported the creation of the first human-pig chimeras by injecting human PSCs into pig blastocysts [38]. The success rate of generating human-pig chimeras was indeed very low, and the chimeras only carried very few human cells, less than one human cell per 100,000 pig cells. To date, no one has reported using gene editing techniques performed in rat embryos to disable the pigs forming a particular organ and enable the human cells to develop more humanlike organs. Ross et al. demonstrated the generation of sheep-human hybrids, and as a step further than the reported human-pig chimeras, the contribution of human cells in sheep embryos was increased to one in 10,000 sheep cells [39]. Although these studies represent only a preliminary step toward the long-term goal, the results from these studies suggest that such creation may be eventually used to grow human organs. As the pigs and sheep are similar in size to humans, the human organs grown in these animals, the heart, liver, kidney, pancreas, lungs, and brain, could be harvested and transplanted into people, meeting the high demand for organ transplantation in the end-stage diseases. In addition, these human-animal chimeras could also be used to investigate the mechanisms of prenatal development and to investigate experimental drugs in different diseases.

Despite all the promises of using human/nonhuman chimeras for regenerative medicine, they also have raised serious ethical dilemmas about the morality of these chimeras. One of the biggest concerns is whether the human cells migrate to the brain and the chimeras end up with a humanlike mind. Such issues could potentially be prevented through genetic editing to avoid the human cell differentiating into the human brain or human gonads. So far, investigations in this field are moving forward with caution, and the reported human/nonhuman chimeras have not been allowed to develop past the fetal stage.

2.6 iPSC banking and allogeneic cell therapies

Although the iPSC cell therapy theoretically enables autologous transplantation, which would eliminate the need for immunosuppression, the inefficiency of iPSC derivation, the time and cost for developing each personalized cell product, and the safety of the products have made such autologous therapies unpractical, particularly for diseases that require an immediate treatment. For the clinical use of iPSC-based cell therapies, it is essential to produce high-quality and safe (no induced mutations in the genome) iPSCs. As will be mentioned below, the pioneering iPSC clinical study in Japan using patients' own iPSC-derived retinal epithelial cells for the treatment of macular degeneration was put on hold due to genomic mutations in the iPSCs. Therefore, the most feasible application of iPSC-based cell therapy would rely on the banked and human leukocyte antigen (HLA)-typed iPSCs, in which the quality and safety have been validated in advance, in the setting of an allogeneic transplantation. This use of allogenic iPSCs however means that immunosuppression would have to be applied to prevent immune rejection. Kawamura et al. recently demonstrated that even though the immunogenicity of allogenic iPSC-derived cardiomyocytes was reduced by major histocompatibility complex (MHC) class I- and class II-matched transplantation in the macaque (monkey), the recipients still required substantial and highly toxic immunosuppression for sustained allogeneic cell engraftment [40]. It has been suggested that the MHC-matched iPSC-derived cardiomyocytes were still susceptible to natural killer (NK) cell destruction, leading to their rejection in the recipients in the absence of immunosuppression [40]. Forced expression of HLA alpha chain E (HLA-E) in PSCs and their differentiated derivatives has been demonstrated to prevent allogeneic response and lysis by NK cells [41]. Recently, Deuse et al. looked into the expression of genes in syncytiotrophoblast, an interface between fetus and mother, and identified low MHC class I and II expression and a high CD47 expression as the features that are responsible for the immune tolerance of syncytiotrophoblast toward allogenic fetal antigens [42]. CD47 is a membrane protein that interacts with several cell surface receptors to inhibit phagocytosis [43]. Indeed, CD47 is a "don't eat me" signal highly expressed on the surface of cancer cells to escape the innate immune responses [44]. The authors then inactivated MHC class I and II genes through CRISPR-Cas9 targeting and overexpressed CD47 via lentiviral transduction in both human and mouse iPSCs [43]. Importantly, the engineered iPSCs and derivatives (endothelial cells, smooth muscle cells, and cardiomyocytes) lost their immunogenicity and persisted long term in fully MHC-mismatched recipients without the use of immunosuppression [43]. This suggests that hypoimmunogenic cell grafts can be engineered from iPSCs for universal transplantation without immunosuppression. These approaches are associated with potential risks of uncontrollable malignant transformation or impaired immune reactions using hypoimmunogenic cell grafts, and consideration of designing an inducible killing switch in the engineered cells to ensure overall safety should be taken into account.

2.7 Current clinical trials with iPSCs

In 2014, Mandai et al. reported the results of the world's first clinical study of iPSC-based therapy in patients with advanced neovascular age-related macular degeneration (AMD) [45]. In this trial, two patients were recruited, and iPSCs were generated from the skin fibroblasts and were further differentiated into retinal pigment epithelial (RPE) cells [45]. One patient received the autologous iPSC-derived RPE cell sheet under the retina. A one-year follow-up on this patient revealed no apparent improvement nor worsening in her vision, and the transplanted sheet remained intact [45]. For the other patient, however, aberrations in DNA copy number were identified in the derived iPSCs and RPE cells, but not the starting fibroblasts, implying that genome mutations occurred during the reprogramming process [45]. This patient did not receive the treatment. Indeed, this clinical trial was suspended due to the discovery of genetic mutations in the iPSCs. In 2017, five patients were recruited and treated for the same eye condition with iPSC-derived retinal cells. However, in this trial, the iPSCs were of an allogeneic source and created at Kyoto University Center for iPS Cell Research and Application (CiRA). One of the patients developed a serious reaction to the transplant. After removal of the engineered membrane graft, the symptoms were improved [46]. The efficacy of the treatment in other patients has not been reported.

In addition to the iPSC-based therapies in AMD, in May 2018, Japan's health ministry approved cardiac surgeon Yoshiki Sawa at Osaka University to assess the safety of allogeneic iPSC-derived cardiomyocytes in patients with heart disease [47]. This research team has previously reported the efficacy of grafting of human iPSC-derived cardiomyocytes cell sheet in combination with an omental flap technique in a porcine model of ischemic cardiomyopathy [48, 49]. In the projected human trial, the treatment will initially be given to three people; then the team will seek approval to conduct a clinical trial in approximately 10 patients [47]. If these initial clinical studies prove successful, the treatment will be made commercially available soon after under a new fast-track system in Japan designed to speed up the development of regenerative therapies [47].

In October 2018, neurosurgeons at Kyoto University Hospital also performed the first iPSC cell-based therapy in patients with Parkinson's disease (https://www.kyoto-u.ac.jp/en/research/events_news/department/hospital/news/2018/181109_1.html). In this first human study, 2.4 million allogeneic iPSC-derived dopamine precursor cells were deposited into 12 sites of the patient's brain with known dopamine activity. At the time of the press conference on this procedure, November of 2018, the investigators described that the patient was "doing well." The human iPSC-derived progenitor cells have shown to improve the symptoms in a primate model of Parkinson's disease [50].

As compared to Japan's fast-forwarding pace of initiating human trials with iPSC-based therapies, the scientists and physicians in the United States are approaching this direction with more caution, even though several human ES cell-based therapies have been initiated in clinical studies [51]. For the treatment of Parkinson's disease, Lorenz Studer at Memorial Sloan Kettering Cancer Center has focused on generating dopamine neurons from human ES cells at a sufficient scale and purity and demonstrated their efficient engraftment and function in mouse, rat, and monkey models of Parkinson's disease [52]. Based on those results, the group is currently pursuing an investigational new drug (IND) application from the US Food and Drug Administration (FDA), to initiate the first human clinical use of ES cell-derived dopamine neurons [52]. In November 2018, Fate Therapeutics, Inc., a biopharmaceutical company, announced that the FDA approved their IND application for FT500, the company's universal NK cells derived from a clonal master iPSC line. Using an in vitro three-dimensional tumor spheroid model, the company demonstrated that FT500, in combination with activated T cells and an anti-programmed death (PD)-1 antibody, led to near-complete elimination of target cells (>99% reduction) [53]. The company plans to initiate first-in-human clinical testing of FT500 in combination with checkpoint inhibitor therapy for the treatment of advanced solid tumors. This is expected to be the first-ever clinical investigation in the United States of an iPSC-derived cell product.

In 2016, Cynata Therapeutics also launched a phase I clinical trial in both the United Kingdom (UK) and Australia using allogeneic iPSC-derived MSCs (differentiated from iPSCs through intermediate-stage mesenchymoangioblasts) (CYP-001) for the treatment of steroid-resistant acute graft-versus-host disease (aGvHD) in patients undergoing an allogenic stem cell transplantation. In 2018, the company reported that CYP-001 met all clinical endpoints and demonstrated positive safety and efficacy data for the treatment of steroid-resistant aGvHD in a phase 1 trial. Cynata plans to advance the cell product into phase 2 trials for GvHD and critical limb ischemia.

3. Adult (somatic) stem cells

3.1 Tissue-resident adult stem cells

Tissues and organs in our body constantly regenerate throughout our lives. Human red blood cells have a defined life span of 120 days [54]. The lining in the intestine turns over completely about every 7 days [55]. Skin cells are constantly shed and then renewed. Endothelial cells in the human heart have a > 15% turnover rate per year, while cardiomyocyte exchange is highest in early childhood and decreases gradually throughout life to <1% per year in adulthood [56]. These are examples of cell turnover at a steady state. Some organs, such as the liver, display relatively slow tissue turnover at a steady state, however upon damage, undergo bursts of cell proliferation and repair [57]. Essential for such replenishment and repair in post-natal life are the activities of adult stem cells. Adult stem cells are undifferentiated tissue-resident cells capable of self-renewal and differentiation within the tissue or organ. Self-renewal allows replication of themselves to sustain their population, and differentiation enables replenishment of various mature cell types upon tissue remodeling and repair, to maintain tissue homeostasis throughout postnatal life.

Hematopoietic stem cells (HSCs) are the first identified adult stem cells [58, 59]. The idea of HSCs, however, can be traced back before the experimental proof of HSCs and following the atomic bombing of Hiroshima and Nagasaki, when the civilians exposed to irradiation were documented to have symptoms of hematopoietic failure. Subsequent experiments in mice demonstrated recovery of the hematopoietic system after radiation injury by bone marrow transplantation [60]. These studies and subsequent transplantation experiments in mice with acute leukemia [61] led to pioneering of allogeneic hematopoietic stem cell transplantation (alloHSCT) by Thomas et al. in patients treated with radiation and chemotherapy in 1957 [62]. At that time, little was known about HLAs, and there was no matching between donors and recipients, until the mid-late 1960, when methods to identify HLA were developed [63]. While many physicians had doubts on the approach of alloHSCT, Thomas et al. persevered and performed transplantation using a matched sibling donor for a patient with leukemia in 1969 [64, 65]. Eight years later, his team performed the first HLA-matched transplant from an unrelated donor, a success that led to the formation of a national registry of bone marrow donors [63]. Dr. Thomas received the 1990 Nobel Prize in Physiology or Medicine with Dr. Joseph Murray, who performed the first successful kidney transplant, for their contribution to cell and organ transplantation.

For decades, hematopoiesis has been described as a cellular hierarchy where only one type of HSC is sitting at the apex of the hierarchy and giving rise to lineage-restricted progenitors, which further differentiate into all the cell types of the blood [66]. However, accumulating data on single HSC repopulation and serial transplantation have revealed an inherent heterogeneity in self-renewal and multi-lineage differentiation of HSCs [67, 68]. These and other studies have contributed to a change of view on HSCs from a single stem cell-type hierarchy model to a consortium model where a pool of stem cells with slightly different properties regenerate all the blood types [69–71]. However, a recent study suggested that platelet-biased stem cells, primed toward the megakaryocyte lineage, reside at the apex of the HSC hierarchy [72]. To reconcile with different observations on HSCs, a new speculative

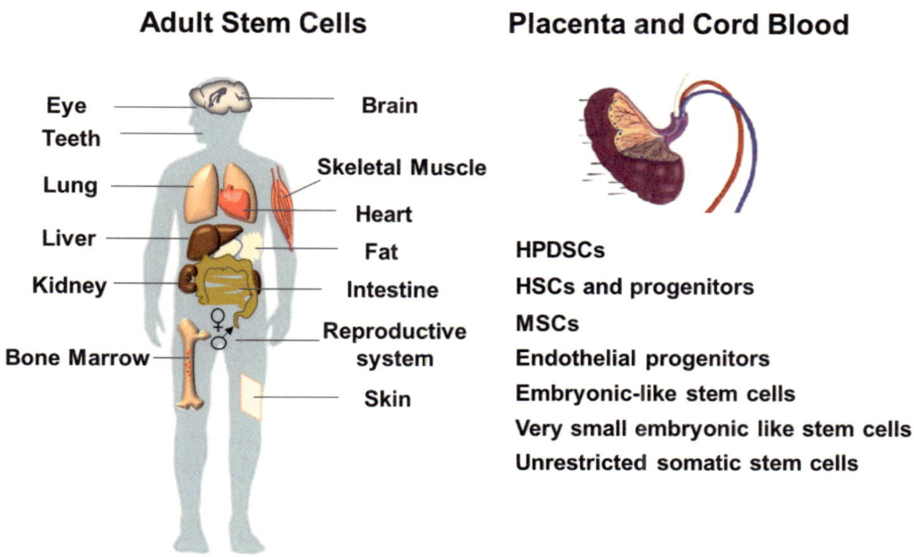

Figure 3.
Identification of adult stem cells from postnatal organs and tissues such as the brain, eye, teeth, lung, heart, kidney, intestine, skeletal muscle, fat tissue, bone marrow, skin, and reproductive system. Human placenta and cord blood are also rich in different hematopoietic stem cells and progenitors, as well as non-hematopoietic stem cells.

model was further raised suggesting that stem cells are rare reserve cells that occasionally generate lineage-restricted progenitors [73]. Some progenitors may retain a degree of developmental flexibility and, during extreme stress, can revert to a stemlike state [73].

Adult stem cells have been identified in most, if not all, tissues and organs in our body, as well as the placenta and cord blood (CB) (**Figure 3**). Extensive studies have been performed to characterize the adult stem cells in the fast-regenerating organs such as the skin and intestine [74, 75]. Moreover, the organs that were once thought to be non-regenerative, such as the brain and heart, also contain adult stem cells, i.e., neural stem cells and cardiac stem cells, respectively [76, 77]. In addition, similar to the evolving concept of hematopoiesis [73], heterogeneity of stem cell populations in distinct compartments in these organs has also been demonstrated, and the heterogeneity enables them to adapt to diverse roles and functions in response to environmental stimulation [73].

3.2 Adult stem cell niche and aging

It has been well accepted that the activities of adult stem cells, whether they stay quiescent or undergo activation, are largely instructed by the microenvironment they reside in, i.e., stem cell niche. Extensive studies have been performed to characterize the stem cell niche in different organ systems such as the skeletal muscle, bone marrow, skin, intestines, and brain (reviewed in [78]). There are common features of the adult stem cell niche. In general, the stem cell niche or microenvironment is composed of stem cells themselves and progeny, surrounding MSCs or stromal cells, extracellular matrix, and adhesion molecules, as well as the external cues from distant sources (long-range signaling factors) [78]. Collectively, both cellular and acellular components of the stem cell niche create a complex microenvironment maintaining stem cell fate and ensuring robust regenerative responses to external stimuli. Although other signaling pathways have been described, TGFβ superfamily, Wnt pathway, and Notch signaling have been identified in different model systems as key regulators of stem cell quiescence, survival, maintenance, and activation [78]. For example, the TGFβ superfamily not only contributes to stem cell quiescence maintenance [79–81], it also plays an important role for stem cell activation and aging. TGFβ has been found to be upregulated in both the satellite cells (skeletal muscle stem cells) and serum of aged mice, which induces high levels of pSmad3 in satellite cells and interferes with their regenerative activities [82]. Importantly, Notch signaling antagonizes pSmad3 and controls satellite cell proliferation by blocking TGFβ-dependent upregulation of cyclin-dependent kinase inhibitors [82]. Meanwhile, increased Wnt signaling in the aged satellite cells has been demonstrated to contribute to cell fate conversion of satellite cells from a myogenic to a fibrogenic state [83].

Recently, Tikhonova et al. mapped the transcriptional landscape of mouse bone marrow microenvironment (HSC niche) at a single-cell resolution and reported previously unappreciated levels of cellular heterogeneity within the niche and with defined distribution of pro-hematopoietic factors [84]. Furthermore, bone marrow niche underwent transcriptional remodeling under stress conditions, leading to a significant upregulation of adipogenesis-related pathways and a global reduction in osteo-lineage-related gene expression [84]. There was also a downregulation of vascular-endothelial-expressed Notch ligand DLL4, which skewed bone marrow hematopoiesis toward a myeloid transcriptional program [84]. These results indeed provided an explanation to the observation that the HSC populations in the elderly exhibit myeloid skewing and lymphoid lineage deficiency [85].

The notion of aging has been raised as the ratio of tissue attrition to tissue regeneration [86]. This process is accompanied by reduced regenerative capacity of adult stem cells at levels of both self-renewal and differentiation (reviewed in [87]). The decline in regenerative function of adult stem cells also contributes to pathophysiological alterations in age-related diseases. Meanwhile, current studies also indicate that age-imposed biochemical changes in the stem cell niche are responsible for such regenerative declines of tissue maintenance and repair. Indeed, several studies have revealed that regenerative potential of stem cells was not controlled by the age of the stem cells themselves, but by the age of the niche they stay in [86]. In a classic experiment, minced skeletal muscle tissue containing satellite cells from young rodents was transplanted into the muscle of aged hosts, and conversely muscle tissue from old rodents was placed into the muscle of young hosts [88]. Such heterochronic transplantation demonstrated that the age of the host was more important than the age of the transplanted stem cells in muscle regeneration [88]. Serial transplantation of spermatogonial stem cells to the testes of young male mice demonstrated that spermatogenesis from that stem cell could continue for more than 3 years, long past the normal life span of the animal, when the stem cell is continually maintained in a young niche [89]. The importance of stem cell niche is further revealed by the heterochronic parabiotic studies [83, 90, 91]. In such experiments, two rodents, young-to-old (heterochronic) and young-to-young and old-to-old (isochronic), are surgically connected through a large flap of the skin, allowing vascular circulation between the two connected animals. In these studies, the isochronic parabionts were not significantly different in tissue regeneration than their respective non-parabiotic age-matched controls. However, in the heterochronic parabionts, regeneration in the muscle, liver, and brain was all significantly improved for the old animal and was decreased for the young animal [83, 90, 91]. As there was no evidence of blood cell exchange between the two connected animals, the studies suggested that tissue regeneration in the stem cells of an old animal could be promoted by "young" systemic factors (long-range factors), while the aged circulation might have "negative" regulators that suppress the regenerative activities of "young" adult stem cells [86]. Examples of identified "negative" factors include p16INK4a, C-C motif chemokine 11 (CCL11), TGFβ, and TNFα, and "young" factors are sirtuin 6 (SIRT6) and Delta/Notch [82, 91–94].

All these studies underscore the importance of understanding how the stem cell microenvironment controls stem cell dynamics and function. Identifying specific molecular and cellular mechanisms that either rejuvenate or compromise the regenerative capacity of adult stem cells will be instructive in developing therapies antagonizing aging and age-related pathological changes. In animal studies, it has been demonstrated that forced activation of Notch signaling and inhibition of TGFβ and Wnt pathways could restore the regenerative capacities of aged muscle [83, 95].

Understanding the impact of microenvironment on stem cell activities is also essential for the development of an effective stem cell therapy. Whether the cells are derived from pluripotent stem cells or isolated from adult tissue, ex vivo culture or manipulation in a microenvironment that maintains cell identity and potency is critical for the efficacy of cell transplantation. Furthermore, a substantial challenge in the regenerative therapy is acute cell death after transplantation of the cells into a degenerative/pathological microenvironment. A recent clinical islet transplantation study quantitated the level of circulating cell-free (cf) DNA as a biomarker for the dead beta cells after transplantation [96]. The authors reported that a distinctive peak of cfDNA was observed 1 hour after transplantation in 83.8% of patients. The cfDNA was also detected 24 hours posttransplant, and that signal was correlated with overall poor clinical outcome (higher insulin requirement, lower stimulated C-peptide level, and decreased 3-month engraftment) [96]. Instant blood-mediated inflammatory response has been speculated to contribute to the significant graft attrition at this early stage of transplant [97]. Therefore, modulation on microenvironment that ensures the survival, integration, and function of the therapeutic cells after transplantation would greatly improve the efficacy of cell therapies.

3.3 Mesenchymal stem cells

MSCs were initially identified as rare non-hematopoietic colony-forming units following plastic adherence of bone marrow cells (reviewed in [98]). Although the original notion of MSCs specifically referred to cells in the bone marrow (also called bone marrow stromal cells, BMSCs), MSCs have been derived from other sources such as cord blood, adipose tissue, and dental pulp. MSCs can be identified by the expression of cell surface markers including CD90, CD73, and CD105 and lack of expression of hematopoietic and endothelial cell markers. MSCs have the capacity to differentiate along mesoderm lineage into osteoblasts, chondrocytes, adipocytes, and fibroblasts. MSCs were also reported to give rise to other mesodermal cell types such as cardiomyocytes and endothelial cells, as well as the cells of other lineages such as neurons and hepatocytes. However, as these claims were mainly based on the expression of the markers and not functional studies, whether MSCs truly have such differentiation capacities require more extensive validation.

There are also controversial results on the age-related changes in MSCs. Some studies reported an age-dependent reduction in the number of MSCs isolated from human bone marrow, while others demonstrated no correlation between MSC numbers and age, even in patients with osteoarthritis [99–102]. Functionally, Sun et al. demonstrated using young and aged mice that although the frequency of MSCs was not significantly different between the young and old bone marrow, the self-renewal and bone formation capacity of old MSCs were significantly compromised as compared to the young MSCs [103]. Moreover, similar to the finding in the satellite cells, exposure of the aged MSCs to a young extracellular matrix rejuvenated these functions of aged MSCs [103]. Recent studies also demonstrated that with age as well as treatment of antidiabetic drugs, MSCs favored differentiation into adipocytes resulting in an increased number of adipocytes and a decreased number of osteoblasts, which may be related to osteoporosis. Downregulation of a transcription factor c-Maf has been identified as the age-related switch in MSC differentiation [104].

3.4 Human umbilical cord blood and placental stem cells

The idea to use human umbilical cord blood arose in the early 1980s, when experiments using cord blood from near-term mice demonstrated hematopoietic reconstitution in lethally irradiated mice [105]. Further, cord blood-derived HSCs were found to have a higher proliferative capacity than those in bone marrow and sufficient doses of HSCs, and hematopoietic progenitor cells were contained in a single collection of cord blood [106]. In 1988 a 5-year-old boy with Fanconi anemia underwent the first cord blood transplantation from his HLA-identical newborn sister [107]. In 1993 the first unrelated cord blood transplantation was performed in a 3-year-old with refractory T-cell acute lymphoblastic leukemia. Since then, related and unrelated cord blood has been increasingly utilized as an alternative source of HSCs [108].

In addition to the HSCs and hematopoietic progenitors, many populations of non-hematopoietic stem cells have been reported in human cord blood, such as MSCs, endothelial progenitor cells [109], embryonic-like stem cells (CBE) [110], very small embryonic-like (VSEL) stem cells [111], multi-lineage progenitor cells (MLPC) [112], and unrestricted somatic stem cells (USSCs) (**Figure 3**) [113]. There have been controversial reports on the identity of VSEL. VSEL stem cells were isolated based on the sorting of $CXCR4^{+}SSEA\text{-}1^{+}Sca\text{-}1^{+}Lineage^{-}CD45^{-}$ cells originally from murine bone marrow cells and subsequently in human CB [111]. Several other groups also reported the isolation of VSEL cells from different adult tissues and implied that VSEL cells were originated during embryonic development and deposited in bone marrow and other organs as dormant precursor cells of adult stem cells (reviewed in [114]). These cells are reported to be very small (3–5 μm), possess large nuclei, express pluripotency markers Oct4 and Nanog, and undergo multi-lineage differentiation. However, the stem cell characteristics of this cell type were challenged by the other group [115].

Among the other cord blood-derived non-hematopoietic stem cells, USSCs are the best characterized stem cell population and represent a novel universal allogenic stem cell source for degenerative diseases. USSCs are generated from full-term CB based on outgrowth of plastic adherent and spindle-shaped colonies in the presence of 30% fetal bovine serum, 10^{-7} M dexamethasone, and 2 mM ultra-glutamine in low glucose DMEM [113]. USSCs possess the ability to differentiate in vitro into the bone, cartilage, adipocytes, hematopoietic cells, and neural cells and in vivo into myocardial cells, Purkinje fibers, and hepatic cells [113]. Although USSCs share cell surface marker with MSCs, they have distinct gene expression, epigenetic signatures, and cytokine profiling [113, 116–120]. USSCs have high proliferation and expansion properties. They can be cultured for more than 20 passages without any spontaneous differentiation or slowing down doubling time [113, 116]. This is a significant advantage of USSCs over MSCs, the doubling time of which dramatically increases after four passages, leading to cell senescence [121]. Significantly, even after 13 passages, the average telomere length of USSCs is 8.6kbp, which is significantly longer than

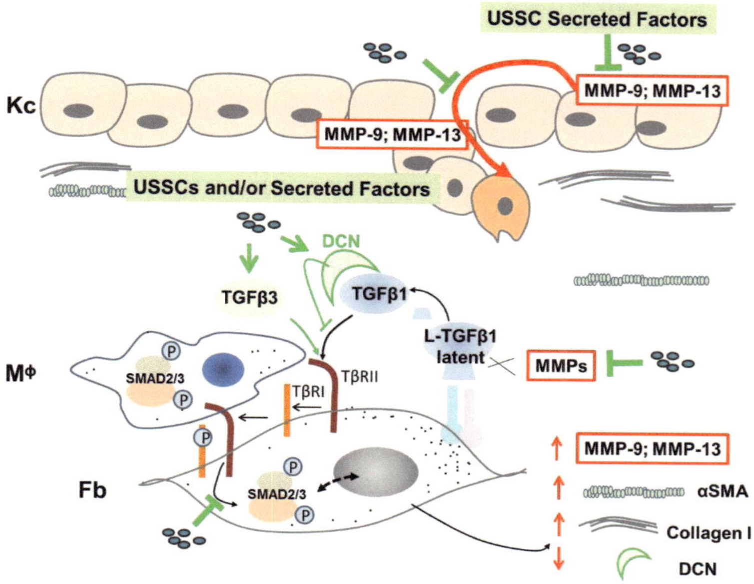

Figure 4.
Dual effects of cord blood-derived USSCs on modulating RDEB skin microenvironment. USSCs exert anti-fibrotic function by suppressing phosphorylation of Smad2/3 in the fibroblasts (Fb) and macrophages (MΦ) of the skin, inhibiting matrix metalloproteinases (MMP)-9 and -13 dermal expression and upregulating anti-fibrotic TGFβ3 and DCN expression. USSCs also attenuate secretion of MMP-9 and MMP-13, which correlate with epithelial malignant transformation, from keratinocytes (Kc) and cutaneous squamous cell carcinoma (cSCCs) derived from patients with RDEB. This figure was adapted from Stem Cells *with permission [125].*

the 7.27kbp telomere length of MSCs at passage 4 [113]. In preclinical studies, USSCs demonstrated the ability to alleviate myocardial infarction, liver injury, spinal cord injury, and intraventricular hemorrhage [122–124]. Moreover, USSCs also promote wound healing and improve manifestation of an inherited skin blistering disease, i.e., RDEB [120, 125]. Suppression of TGFβ-mediated fibrosis and modulation of extracellular matrix remodeling have been accounted as part of the mechanisms of action of USSCs in the treatment of RDEB (**Figure 4**) [125].

Once considered as a medical waste similar to CB, the human placenta has also been demonstrated to provide a novel stem cell source for cellular therapy. Hematopoietic stem and progenitor cells were identified throughout gestation from human placental blood, vessel perfusate, and cells from digested placenta tissue, and stromal cells generated from placenta possessed pericyte characteristic and may be a supportive microenvironment for hematopoiesis [126]. Recently Celgene Cellular Therapeutics, Inc. manufactured human placental-derived stem cells (HPDSCs) from full-term donor placentas following saline perfusion, red blood cell depletion, and volume reduction. The overall cell types as determined by flow cytometry analysis are similar between HPDSCs and CB [127]. However, HPDSCs contain a significantly higher level of both hematopoietic and non-hematopoietic stem and progenitor cells than CB. In addition, HPDSCs have a lower percentage of T cells than CB and are largely negative for MHC class II molecules, indicative of their potential use as both autologous and allogeneic cells. A pilot clinical study demonstrated that adding HPDSCs as universal donor cells with CB transplantation in patients with malignant and nonmalignant diseases had no adverse effects and may reduce the incidence of aGvHD [128]. Administration of HPDSCs in a mouse model of RDEB, in the absence of any conditioning regimen, also resulted in significant improvement on the survival and disease manifestation of the recipient mice [127].

As will be mentioned below, CB hematopoietic progenitor cells are among the few stem cell products that are approved by the FDA. Over 40,000 CB transplantations have been performed worldwide in both adults and children for the treatment of around 80 different disorders [129]. The advantage of using CB and cord blood CB-derived stem cells compared to other adult stem cell sources is the fast availability and ease in collection without causing any discomfort or risk to the donors. Moreover, being early in development, CB stem cells have not been exposed to immunological challenge and are less likely to carry somatic mutations than other adult cells. Any age- or stress-related transcriptional remodeling that might have impacted the stem cell function of adult stem cells, as discussed above, would not be of an issue in CB stem cells.

3.5 Examples of clinical studies with adult stem cells

As of today, FDA-approved cellular and gene therapy products are limited to the hematopoietic progenitor cells from CB for alloHSCT in patients with disorders affecting the hematopoietic systems, autologous chondrocytes on a porcine collagen membrane for repair of cartilage defects of the knee, allogeneic cultured keratinocytes and fibroblasts in the treatment of mucogingival conditions, and a few chimeric antigen receptor (CAR) autologous cellular immunotherapies (https://www.fda.gov/vaccines-blood-biologics/cellular-gene-therapy-products/approved-cellular-and-gene-therapy-products).

In addition to the treatment of the disorders affecting hematopoietic system, CB alloHSCT has also been applied in clinical trials to treat nonmalignant diseases. More than 20 years ago, CB transplantation had been initiated in infants and children with Krabbe's disease [130, 131]. Krabbe's disease is an autosomal recessive disorder due to deficiency of the lysosomal enzyme galactocerebrosidase, leading to progressive neurologic deterioration and death in early childhood. These transplantation studies demonstrated that CB cells, both hematopoietic and non-hematopoietic in origin, could engraft in the patients' central nervous system, providing the missing enzyme and facilitating neural cell repair. Particularly, the young patients that underwent transplantation before the development of symptoms showed significant improvements in developmental skills, while the children who underwent transplantation after the onset of symptoms had minimal neurologic improvement [130, 131]. A long-term follow-up (median 9.5 years, range 4–15 years) study further demonstrated that the surviving patients who underwent early CB transplantation function at a much higher level than untreated children or children who were symptomatic at the time of alloHSCT [132]. Based on the observed efficacy of CB transplantation on improving neurological outcome, Cotten et al. conducted a pilot study on intravenous infusion of autologous CB in 184 pediatric patients who had their CB banked at birth and were subsequently diagnosed with acquired neurological disorders [133]. This investigation demonstrated the safety and feasibility of autologous CB infusion in these pediatric recipients. Subsequently, Cotten et al. also reported the safety and feasibility of autologous CB infusion in neonates diagnosed as hypoxic ischemic encephalopathy [134].

Over the last two decades, there have been an increasing number of clinical trials involving stem cell treatment for various degenerative diseases. For cardiovascular disease alone, there have been more than 200 clinical trials using various cell types including skeletal myoblasts, autologous bone marrow mononuclear cells (BMMNCs), $CD133^+$ bone marrow cells, endothelial progenitors, autologous $CD34^+$ cells, MSCs, cardiopoietic stem cells that were generated by treating MSCs with a cocktail of trophic factors, allogeneic and autologous c-kit$^+$ cardiac stem cells isolated from biopsies obtained during coronary artery bypass grafting, and cardiosphere-derived cells generated from culture outgrowth of heart biopsies (reviewed in [135]). Most studies demonstrated that the tested cells were safe and the procedures were feasible. However, the efficacy of these cell therapies remains inconclusive. Early clinical trials, ranging from pilot to phase III double-blinded placebo-controlled studies, such as TOPCARE-AMI, BOOST, FNCELL, and REPAIR-AMI, demonstrated significant efficacy (evaluated-based on left ventricular ejection fraction, LVEF) of BMMNC treatment in patients with acute myocardial infarction. However, recent multicenter, double-blinded, and placebo-controlled studies, such as TIME, LateTIME, and SWISS AMI trials, found no improvements after BMMNC administration in LVEF and other parameters measured by cMRI. Even the same center, which reported significant improvement in heart function in the initial pilot trial and subsequent partly randomized and open-label study on autologous CD133$^+$ bone marrow cell infusion during coronary artery bypass graft surgery [136, 137], later revealed no significant benefits of CD133$^+$ bone marrow cell infusion in a strictly double-blind, fully randomized, and placebo-controlled trial (CARDIO133, NCT00462774) [138]. It has also to be mentioned that a report on the initial results of a phase I, open-label, and randomized trial using autologous c-kit+ cardiac stem cells in patients with ischemic cardiomyopathy was recently retracted from the journal, due to lack of reliability on the production of cells and integrity of the data [139].

3.6 Lessons learned from clinical stem cell therapies

Although there are a substantial number of clinical studies, it is difficult to compare results between trials. Sources of cell products, mode of cell delivery and cell dose, timing of administration, the age and complications of patients, and the choice of surrogate endpoint markers are all the variables for the efficacy of each clinical study.

When it comes to a stem cell therapy, it is tempting to think of it as the cell replacement therapy, that is, the stem cells engraft and differentiate to replace the damaged cells in vivo. However, although a few studies demonstrated that MSCs could undergo tri-lineage differentiation and human CD34$^+$ cells could fuse with cardiomyocytes and also transdifferentiate in animal models [140, 141], most studies reveal no evidence of therapeutic cells to undergo real cell replacement in vivo. Even c-kit$^+$ cardiac stem cells and cardiosphere-derived cells do not become cardiomyocytes in experimental animals with infarction [142, 143]. Rather, paracrine effects, e.g., secretion of growth factors to enhance tissue preservation and/or recruitment of endogenous repair, have been considered as the major mode of action of stem cells [137,138]. Similar lack of differentiation has also been reported when neural stem cells or progenitors were transplanted in animal models [144]. It would be interesting to see whether administration of human ES or iPSC-derived mature cells, such as cardiomyocytes and neurons as mentioned in the first part of the chapter, would result in more effective outcomes in the recruited patients, as in preclinical studies, these fully differentiated cells appeared to be able to functionally integrate in the recipient tissue [49, 50].

For cellular therapies, there is a choice of using either autologous or allogeneic cells. The autologous cells have the advantage of circumventing immunogenicity; however, the cell dose could be limited per isolation and the quality may not be reliable. Moreover, as mentioned above, the function of tissue stem cells declines with age. It was also reported that BMMNCs isolated from patients with chronic ischemic heart disease have a significantly reduced migratory and colony-forming activity in vitro and a reduced neovascularization capacity in vivo, as compared to healthy controls [145]. Therefore, for the patients that are elderly and suffer from pathological insults, they may not benefit from treatment using their own cells. For allogeneic cells, the cell dose is less of an issue than autologous cells. As off-the-shelf products, they are presumably validated for safety and efficacy before being applied to patients. However, generation of cell lots with consistent potency and comparability has been recognized as a significant issue for clinical translation. Recently, two independent investigations reported that clinical-grade human neural stem cell product (HuCNS-SC; proprietary of StemCells, Inc.), in contrast to the research-grade NSCs provided by the same company, failed to demonstrate the efficacy in animal models of spinal cord injury and Alzheimer's disease, respectively [144, 146]. However, despite being informed of the negative impact of stem cell engraftment on functional outcome in the animal model, the company initiated a clinical trial testing this product in patients with cervical spinal cord injury (NCT02163876) in December 2014 and subsequently reported a small improvement in motor strength in 4/5 subjects in the 6-month interim report. However, the clinical trial was terminated in May 2016, due to a lack of significant improvements and the lack of a trend for improvements over time [146]. In this study, there were no details on how the clinical- and research-grade products were made differently that might contribute to the disparity on the efficacy of animal studies. HuCNS-SC was derived from donated fetal brain tissue based on fluorescence-activated cell sorting of $CD133^+$ cells and expanded as neurospheres. It is possible that the clinical- and research-grade products were from different donors with varying genetic background and/or developmental stage. It is also possible that the scale-up production of the cell products under good manufacturing practice (GMP) unfavorably changed the therapeutic function of the cells. Nevertheless, the lack of efficacy of the cell product in the animal model was consistent with the failure of the clinical trial. It has to be noted that in vivo preclinical testing of the final clinical product is not required by the FDA, because "human-derived cellular therapy products intended for clinical administration in animals may not be informative" [147]. As a result, stem cell products are increasingly entering clinical studies for various disease conditions without prior efficacy studies in animal models, outpacing our understanding on their potential mechanisms of action. Although it is true that not all the animal models recapitulate spectrum of human diseases and it is difficult to extrapolate the results from the animal study to human, the lesson learned from the failures of clinical studies including HuCNS-SC is the importance of robust and reliable potency assays to characterize the cells and to ensure the consistency between different manufacturing lots before applying to patients. Recently, the International Society of Stem Cell Research (ISSCR) released updated guidelines on stem cell research and clinical translation, recommending that the cells entering clinical trials are based on sound scientific rationales with robust manufacturing and animal efficacy data, in addition to a safety package to support clinical trials [148].

4. Conclusion

We are at the end of the beginning in the field of stem cell research and stem cell regenerative medicine. Particularly, the development of iPSCs and gene editing techniques have opened a new era of disease modeling and personalized medicine. However, although it is feasible, the time and cost for production and validation of autologous iPSC-derived cellular therapeutics have made such personalized medicine unpractical, especially for diseases requiring an immediate therapeutic intervention. Therefore, a more feasible use of iPSCs and iPSC-derived cell products would rely on an HLA-matched allogeneic setting. Moreover, targeted differentiation of PSCs into distinct cell subpopulations and proper cell maturation remain to be a challenge in the development of iPSC therapies. In the end, whether the cell products are derived from iPSCs or adult stem cells, they face the same challenges as cellular therapies, that is, large number of cell death after transplantation and poor functional integration of the survived cells. The current efforts on tissue engineering and organoid system have demonstrated promises in overcoming these difficulties. Understanding tissue microenvironment is also the key to develop effective therapies that ensure exogenous cell engraftment and integration and/or augment endogenous tissue stem cell function for regeneration.

With the excitement of stem cell research, more subtypes of stem cells are entering into clinical studies, and there is a growing interest in commercializing and marketing of these stem cell products. However, caution should be maintained to ensure the quality of cell products and the scientific rationale and rigor for their clinical translation. Lastly, a randomized and controlled clinical trial with large sample size and multiple surrogate endpoints are essential to determine the safety and efficacy of stem cell therapy.

Acknowledgements

The authors would like to thank Erin Morris, RN, for assistance in the preparation of this chapter. This work was supported by grants from the Pediatric Cancer Research Foundation to MSC, DEBRA International funding to MSC, and National Natural Science Foundation of China (81472141) to HZ.

Author details

Yanling Liao[1], Hongwen Zhu[2,3], Larisa Ivanova[1] and Mitchell S. Cairo[1,4,5,6,7]*

1 Department of Pediatrics, New York Medical College, Valhalla, NY, USA

2 Tianjin Hospital, Tianjin Academy of Integrative Medicine, Tianjin, China

3 Medical Technology Research Institute, BOE, Beijing, China

4 Department of Medicine, New York Medical College, Valhalla, NY, USA

5 Department of Pathology, New York Medical College, Valhalla, NY, USA

6 Department of Immunology and Microbiology, New York Medical College, Valhalla, NY, USA

7 Department of Cell Biology and Anatomy, New York Medical College, Valhalla, NY, USA

*Address all correspondence to: mitchell_cairo@nymc.edu

References

[1] Haeckel E. Natürliche Schöpfungsgeschichte. Berlin: Georg Reimer; 1868

[2] Haeckel E. In: Wilhelm Engelmann L, editor. Anthropogenie oder Entwickelungsgeschichte des Menschen. 3rd ed. 1877

[3] Evans MJ, Kaufman MH. Establishment in culture of pluripotential cells from mouse embryos. Nature. 1981;**292**(5819):154-156

[4] Thomson JA, Itskovitz-Eldor J, Shapiro SS, Waknitz MA, Swiergiel JJ, Marshall VS, et al. Embryonic stem cell lines derived from human blastocysts. Science. 1998;**282**(5391):1145-1147

[5] Lo B, Zettler P, Cedars MI, Gates E, Kriegstein AR, Oberman M, et al. A new era in the ethics of human embryonic stem cell research. Stem Cells. 2005;**23**(10):1454-1459. DOI: 10.1634/stemcells.2005-ed.4

[6] Grinnemo KH, Kumagai-Braesch M, Mansson-Broberg A, Skottman H, Hao X, Siddiqui A, et al. Human embryonic stem cells are immunogenic in allogeneic and xenogeneic settings. Reproductive Biomedicine Online. 2006;**13**(5):712-724

[7] Gurdon JB. The developmental capacity of nuclei taken from intestinal epithelium cells of feeding tadpoles. Journal of Embryology and Experimental Morphology. 1962;**10**:622-640

[8] Campbell KH, McWhir J, Ritchie WA, Wilmut I. Sheep cloned by nuclear transfer from a cultured cell line. Nature. 1996;**380**(6569):64-66. DOI: 10.1038/380064a0

[9] Takahashi K, Yamanaka S. Induction of pluripotent stem cells from mouse embryonic and adult fibroblast cultures by defined factors. Cell. 2006;**126**(4):663-676. DOI: 10.1016/j.cell.2006.07.024

[10] Malik N, Rao MS. A review of the methods for human iPSC derivation. Methods in Molecular Biology. 2013;**997**:23-33. DOI: 10.1007/978-1-62703-348-0_3

[11] Loh YH, Agarwal S, Park IH, Urbach A, Huo H, Heffner GC, et al. Generation of induced pluripotent stem cells from human blood. Blood. 2009;**113**(22):5476-5479. DOI: 10.1182/blood-2009-02-204800

[12] Ma H, Morey R, O'Neil RC, He Y, Daughtry B, Schultz MD, et al. Abnormalities in human pluripotent cells due to reprogramming mechanisms. Nature. 2014;**511**(7508):177-183. DOI: 10.1038/nature13551

[13] Johannesson B, Sagi I, Gore A, Paull D, Yamada M, Golan-Lev T, et al. Comparable frequencies of coding mutations and loss of imprinting in human pluripotent cells derived by nuclear transfer and defined factors. Cell Stem Cell. 2014;**15**(5):634-642. DOI: 10.1016/j.stem.2014.10.002

[14] Wolf DP, Morey R, Kang E, Ma H, Hayama T, Laurent LC, et al. Concise review: Embryonic stem cells derived by somatic cell nuclear transfer: A horse in the race? Stem Cells. 2017;**35**(1):26-34. DOI: 10.1002/stem.2496

[15] Shi Y, Inoue H, Wu JC, Yamanaka S. Induced pluripotent stem cell technology: A decade of progress. Nature Reviews. Drug Discovery. 2017;**16**(2):115-130. DOI: 10.1038/nrd.2016.245

[16] Kim C. iPSC technology—Powerful hand for disease modeling and therapeutic screen. BMB Reports. 2015;**48**(5):256-265

[17] Itoh M, Kiuru M, Cairo MS, Christiano AM. Generation of keratinocytes from normal and recessive dystrophic epidermolysis bullosa-induced pluripotent stem cells. Proceedings of the National Academy of Sciences of the United States of America. 2011;**108**(21):8797-8802. DOI: 10.1073/pnas.1100332108

[18] Sances S, Bruijn LI, Chandran S, Eggan K, Ho R, Klim JR, et al. Modeling ALS with motor neurons derived from human induced pluripotent stem cells. Nature Neuroscience. 2016;**19**(4):542-553. DOI: 10.1038/nn.4273

[19] Dimos JT, Rodolfa KT, Niakan KK, Weisenthal LM, Mitsumoto H, Chung W, et al. Induced pluripotent stem cells generated from patients with ALS can be differentiated into motor neurons. Science. 2008;**321**(5893):1218-1221. DOI: 10.1126/science.1158799

[20] Kiskinis E, Sandoe J, Williams LA, Boulting GL, Moccia R, Wainger BJ, et al. Pathways disrupted in human ALS motor neurons identified through genetic correction of mutant SOD1. Cell Stem Cell. 2014;**14**(6):781-795. DOI: 10.1016/j.stem.2014.03.004

[21] Chen H, Qian K, Du Z, Cao J, Petersen A, Liu H, et al. Modeling ALS with iPSCs reveals that mutant SOD1 misregulates neurofilament balance in motor neurons. Cell Stem Cell. 2014;**14**(6):796-809. DOI: 10.1016/j.stem.2014.02.004

[22] Wainger BJ, Kiskinis E, Mellin C, Wiskow O, Han SS, Sandoe J, et al. Intrinsic membrane hyperexcitability of amyotrophic lateral sclerosis patient-derived motor neurons. Cell Reports. 2014;7(1):1-11. DOI: 10.1016/j.celrep.2014.03.019

[23] Matus S, Medinas DB, Hetz C. Common ground: Stem cell approaches find shared pathways underlying ALS. Cell Stem Cell. 2014;**14**(6):697-699. DOI: 10.1016/j.stem.2014.05.001

[24] Sadegh C, Macklis JD. Established monolayer differentiation of mouse embryonic stem cells generates heterogeneous neocortical-like neurons stalled at a stage equivalent to midcorticogenesis. The Journal of Comparative Neurology. 2014;**522**(12):2691-2706. DOI: 10.1002/cne.23576

[25] Focosi D, Amabile G. Induced pluripotent stem cell-derived red blood cells and platelet concentrates: From bench to bedside. Cells. 2017;7(1). DOI: 10.3390/cells7010002

[26] Lu SJ, Feng Q, Park JS, Vida L, Lee BS, Strausbauch M, et al. Biologic properties and enucleation of red blood cells from human embryonic stem cells. Blood. 2008;**112**(12):4475-4484. DOI: 10.1182/blood-2008-05-157198

[27] Chang CJ, Mitra K, Koya M, Velho M, Desprat R, Lenz J, et al. Production of embryonic and fetal-like red blood cells from human induced pluripotent stem cells. PLoS One. 2011;**6**(10):e25761. DOI: 10.1371/journal.pone.0025761

[28] Tallack MR, Perkins AC. KLF1 directly coordinates almost all aspects of terminal erythroid differentiation. IUBMB Life. 2010;**62**(12):886-890. DOI: 10.1002/iub.404

[29] Yang CT, Ma R, Axton RA, Jackson M, Taylor AH, Fidanza A, et al. Activation of KLF1 enhances the differentiation and maturation of red blood cells from human pluripotent stem cells. Stem Cells. 2017;**35**(4):886-897. DOI: 10.1002/stem.2562

[30] Takebe T, Sekine K, Enomura M, Koike H, Kimura M, Ogaeri T, et al. Vascularized and functional human liver from an iPSC-derived organ bud transplant. Nature.

2013;**499**(7459):481-484. DOI: 10.1038/nature12271

[31] Lian X, Hsiao C, Wilson G, Zhu K, Hazeltine LB, Azarin SM, et al. Robust cardiomyocyte differentiation from human pluripotent stem cells via temporal modulation of canonical Wnt signaling. Proceedings of the National Academy of Sciences of the United States of America. 2012;**109**(27):E1848-E1857. DOI: 10.1073/pnas.1200250109

[32] Patsch C, Challet-Meylan L, Thoma EC, Urich E, Heckel T, O'Sullivan JF, et al. Generation of vascular endothelial and smooth muscle cells from human pluripotent stem cells. Nature Cell Biology. 2015;**17**(8):994-1003. DOI: 10.1038/ncb3205

[33] Besser RR, Ishahak M, Mayo V, Carbonero D, Claure I, Agarwal A. Engineered microenvironments for maturation of stem cell derived cardiac myocytes. Theranostics. 2018;**8**(1):124-140. DOI: 10.7150/thno.19441

[34] Truskey GA. Advancing cardiovascular tissue engineering. F1000Research. 2016;**5**. DOI: 10.12688/f1000research.8237.1

[35] Murphy SV, Atala A. 3D bioprinting of tissues and organs. Nature Biotechnology. 2014;**32**(8):773-785. DOI: 10.1038/nbt.2958

[36] Nadav Noor AS, Edri R, Gal I, Wertheim L, Dvir T. 3D printing of personalized thick and perfusable cardiac patches and hearts. Advanced Science. 2019:1900344. DOI: 10.1002/advs.201900344

[37] Yamaguchi T, Sato H, Kato-Itoh M, Goto T, Hara H, Sanbo M, et al. Interspecies organogenesis generates autologous functional islets. Nature. 2017;**542**(7640):191-196. DOI: 10.1038/nature21070

[38] Wu J, Platero-Luengo A, Sakurai M, Sugawara A, Gil MA, Yamauchi T, et al. Interspecies chimerism with mammalian pluripotent stem cells. Cell. 2017;**168**(3):473-486 e15. DOI: 10.1016/j.cell.2016.12.036

[39] Ross P. Towards xenogeneic generation of human organs. In: AAAS Annual Meeting; Austin, TX. 2018

[40] Kawamura T, Miyagawa S, Fukushima S, Maeda A, Kashiyama N, Kawamura A, et al. Cardiomyocytes derived from MHC-homozygous induced pluripotent stem cells exhibit reduced allogeneic immunogenicity in MHC-matched non-human primates. Stem Cell Reports. 2016;**6**(3):312-320. DOI: 10.1016/j.stemcr.2016.01.012

[41] Gornalusse GG, Hirata RK, Funk SE, Riolobos L, Lopes VS, Manske G, et al. HLA-E-expressing pluripotent stem cells escape allogeneic responses and lysis by NK cells. Nature Biotechnology. 2017;**35**(8):765-772. DOI: 10.1038/nbt.3860

[42] Deuse T, Hu X, Gravina A, Wang D, Tediashvili G, De C, et al. Hypoimmunogenic derivatives of induced pluripotent stem cells evade immune rejection in fully immunocompetent allogeneic recipients. Nature Biotechnology. 2019;**37**(3):252-258. DOI: 10.1038/s41587-019-0016-3

[43] Jaiswal S, Jamieson CH, Pang WW, Park CY, Chao MP, Majeti R, et al. CD47 is upregulated on circulating hematopoietic stem cells and leukemia cells to avoid phagocytosis. Cell. 2009;**138**(2):271-285. DOI: 10.1016/j.cell.2009.05.046

[44] Majeti R, Chao MP, Alizadeh AA, Pang WW, Jaiswal S, Gibbs KD Jr, et al. CD47 is an adverse prognostic factor and therapeutic antibody target on human acute myeloid leukemia stem cells. Cell. 2009;**138**(2):286-299. DOI: 10.1016/j.cell.2009.05.045

[45] Mandai M, Watanabe A, Kurimoto Y, Hirami Y, Morinaga C, Daimon T, et al. Autologous induced stem-cell-derived retinal cells for macular degeneration. The New England Journal of Medicine. 2017;**376**(11):1038-1046. DOI: 10.1056/NEJMoa1608368

[46] Kyodo. First serious adverse reaction to iPS-derived retinal cell transplant reported. The Japan Times. 2018

[47] Cyranoski D. 'Reprogrammed' stem cells approved to mend human hearts for the first time. Nature. 2018;**557**(7707):619-620. DOI: 10.1038/d41586-018-05278-8

[48] Kawamura M, Miyagawa S, Fukushima S, Saito A, Miki K, Funakoshi S, et al. Enhanced therapeutic effects of human iPS cell derived-cardiomyocyte by combined cell-sheets with omental flap technique in porcine ischemic cardiomyopathy model. Scientific Reports. 2017;**7**(1):8824. DOI: 10.1038/s41598-017-08869-z

[49] Ishida M, Miyagawa S, Saito A, Fukushima S, Harada A, Ito E, et al. Transplantation of human-induced pluripotent stem cell-derived cardiomyocytes is superior to somatic stem cell therapy for restoring cardiac function and oxygen consumption in a porcine model of myocardial infarction. Transplantation. 2019;**103**(2):291-298. DOI: 10.1097/TP.0000000000002384

[50] Kikuchi T, Morizane A, Doi D, Magotani H, Onoe H, Hayashi T, et al. Human iPS cell-derived dopaminergic neurons function in a primate Parkinson's disease model. Nature. 2017;**548**(7669):592-596. DOI: 10.1038/nature23664

[51] Martin U. Therapeutic application of pluripotent stem cells: Challenges and risks. Frontiers in Medicine. 2017;**4**:229. DOI: 10.3389/fmed.2017.00229

[52] Studer L. Towards a pluripotent stem cell based therapy for Parkinson's disease. In: ISSCR; Melbourne, Australia. 2018

[53] Miller JS. iPSC-derived NK cells and anti-PD1 antibody synergize to enhance T-cell cytokine and cytolytic responses against multiple tumors ASH. In: San Diego Convention Center. 2018

[54] Pantaleo A, Giribaldi G, Mannu F, Arese P, Turrini F. Naturally occurring anti-band 3 antibodies and red blood cell removal under physiological and pathological conditions. Autoimmunity Reviews. 2008;**7**(6):457-462. DOI: 10.1016/j.autrev.2008.03.017

[55] Atala A. Resident stem cells and regenerative therapy foreword. In: Resident Stem Cells and Regenerative Therapy. 2013. pp. Xi-Xi

[56] Bergmann O, Zdunek S, Felker A, Salehpour M, Alkass K, Bernard S, et al. Dynamics of cell generation and turnover in the human heart. Cell. 2015;**161**(7):1566-1575. DOI: 10.1016/j.cell.2015.05.026

[57] Duncan AW, Dorrell C, Grompe M. Stem cells and liver regeneration. Gastroenterology. 2009;**137**(2):466-481. DOI: 10.1053/j.gastro.2009.05.044

[58] Till JE, Mc CE. A direct measurement of the radiation sensitivity of normal mouse bone marrow cells. Radiation Research. 1961;**14**:213-222

[59] Becker AJ, Mc CE, Till JE. Cytological demonstration of the clonal nature of spleen colonies derived from transplanted mouse marrow cells. Nature. 1963;**197**:452-454

[60] Lorenz E, Uphoff D, Reid TR, Shelton E. Modification of irradiation injury in mice and guinea pigs by bone marrow injections. Journal of the National Cancer Institute. 1951;**12**(1):197-201

[61] Barnes DW, Corp MJ, Loutit JF, Neal FE. Treatment of murine leukaemia with X rays and homologous bone marrow; preliminary communication. British Medical Journal. 1956;**2**(4993):626-627. DOI: 10.1136/bmj.2.4993.626

[62] Thomas ED, Lochte HL Jr, Lu WC, Ferrebee JW. Intravenous infusion of bone marrow in patients receiving radiation and chemotherapy. The New England Journal of Medicine. 1957;**257**(11):491-496. DOI: 10.1056/NEJM195709122571102

[63] van Rood JJ. The detection of transplantation antigens in leukocytes. Seminars in Hematology. 1968;**5**(2):187-214

[64] Thomas ED, Buckner CD, Banaji M, Clift RA, Fefer A, Flournoy N, et al. One hundred patients with acute leukemia treated by chemotherapy, total body irradiation, and allogeneic marrow transplantation. Blood. 1977;**49**(4):511-533

[65] Thomas ED, Buckner CD, Clift RA, Fefer A, Johnson FL, Neiman PE, et al. Marrow transplantation for acute nonlymphoblastic leukemia in first remission. The New England Journal of Medicine. 1979;**301**(11):597-599. DOI: 10.1056/NEJM197909133011109

[66] Spangrude GJ, Smith L, Uchida N, Ikuta K, Heimfeld S, Friedman J, et al. Mouse hematopoietic stem cells. Blood. 1991;**78**(6):1395-1402

[67] Dykstra B, Kent D, Bowie M, McCaffrey L, Hamilton M, Lyons K, et al. Long-term propagation of distinct hematopoietic differentiation programs in vivo. Cell Stem Cell. 2007;**1**(2):218-229. DOI: 10.1016/j.stem.2007.05.015

[68] Notta F, Zandi S, Takayama N, Dobson S, Gan OI, Wilson G, et al. Distinct routes of lineage development reshape the human blood hierarchy across ontogeny. Science. 2016;**351**(6269):aab2116. DOI: 10.1126/science.aab2116

[69] Challen GA, Boles NC, Chambers SM, Goodell MA. Distinct hematopoietic stem cell subtypes are differentially regulated by TGF-beta1. Cell Stem Cell. 2010;**6**(3):265-278. DOI: 10.1016/j.stem.2010.02.002

[70] Morita Y, Ema H, Nakauchi H. Heterogeneity and hierarchy within the most primitive hematopoietic stem cell compartment. The Journal of Experimental Medicine. 2010;**207**(6):1173-1182. DOI: 10.1084/jem.20091318

[71] Oguro H, Ding L, Morrison SJ. SLAM family markers resolve functionally distinct subpopulations of hematopoietic stem cells and multipotent progenitors. Cell Stem Cell. 2013;**13**(1):102-116. DOI: 10.1016/j.stem.2013.05.014

[72] Sanjuan-Pla A, Macaulay IC, Jensen CT, Woll PS, Luis TC, Mead A, et al. Platelet-biased stem cells reside at the apex of the haematopoietic stem-cell hierarchy. Nature. 2013;**502**(7470):232-236. DOI: 10.1038/nature12495

[73] Goodell MA, Nguyen H, Shroyer N. Somatic stem cell heterogeneity: Diversity in the blood, skin and intestinal stem cell compartments. Nature Reviews Molecular Cell Biology. 2015;**16**(5):299-309. DOI: 10.1038/nrm3980

[74] Fuchs E, Horsley V. More than one way to skin. Genes & Development. 2008;**22**(8):976-985. DOI: 10.1101/gad.1645908

[75] Yan KS, Chia LA, Li X, Ootani A, Su J, Lee JY, et al. The intestinal stem cell markers Bmi1 and Lgr5 identify two functionally distinct populations. Proceedings of the National Academy of Sciences of the United States of

America. 2012;**109**(2):466-471. DOI: 10.1073/pnas.1118857109

[76] Fuentealba LC, Obernier K, Alvarez-Buylla A. Adult neural stem cells bridge their niche. Cell Stem Cell. 2012;**10**(6):698-708. DOI: 10.1016/j.stem.2012.05.012

[77] Beltrami AP, Barlucchi L, Torella D, Baker M, Limana F, Chimenti S, et al. Adult cardiac stem cells are multipotent and support myocardial regeneration. Cell. 2003;**114**(6):763-776

[78] Rezza A, Sennett R, Rendl M. Adult stem cell niches: Cellular and molecular components. Current Topics in Developmental Biology. 2014;**107**:333-372. DOI: 10.1016/B978-0-12-416022-4.00012-3

[79] Yamazaki S, Iwama A, Takayanagi S, Eto K, Ema H, Nakauchi H. TGF-beta as a candidate bone marrow niche signal to induce hematopoietic stem cell hibernation. Blood. 2009;**113**(6):1250-1256. DOI: 10.1182/blood-2008-04-146480

[80] Kobielak K, Stokes N, de la Cruz J, Polak L, Fuchs E. Loss of a quiescent niche but not follicle stem cells in the absence of bone morphogenetic protein signaling. Proceedings of the National Academy of Sciences of the United States of America. 2007;**104**(24):10063-10068. DOI: 10.1073/pnas.0703004104

[81] Mira H, Andreu Z, Suh H, Lie DC, Jessberger S, Consiglio A, et al. Signaling through BMPR-IA regulates quiescence and long-term activity of neural stem cells in the adult hippocampus. Cell Stem cell. 2010;7(1):78-89. DOI: 10.1016/j.stem.2010.04.016

[82] Carlson ME, Hsu M, Conboy IM. Imbalance between pSmad3 and Notch induces CDK inhibitors in old muscle stem cells. Nature. 2008;**454**(7203):528-532. DOI: 10.1038/nature07034

[83] Brack AS, Conboy MJ, Roy S, Lee M, Kuo CJ, Keller C, et al. Increased Wnt signaling during aging alters muscle stem cell fate and increases fibrosis. Science. 2007;**317**(5839):807-810. DOI: 10.1126/science.1144090

[84] Tikhonova AN, Dolgalev I, Hu H, Sivaraj KK, Hoxha E, Cuesta-Dominguez A, et al. The bone marrow microenvironment at single-cell resolution. Nature. 2019;**569**(7755):222-228. DOI: 10.1038/s41586-019-1104-8

[85] Cho RH, Sieburg HB, Muller-Sieburg CE. A new mechanism for the aging of hematopoietic stem cells: Aging changes the clonal composition of the stem cell compartment but not individual stem cells. Blood. 2008;**111**(12):5553-5561. DOI: 10.1182/blood-2007-11-123547

[86] Conboy IM, Rando TA. Heterochronic parabiosis for the study of the effects of aging on stem cells and their niches. Cell Cycle. 2012;**11**(12):2260-2267. DOI: 10.4161/cc.20437

[87] Raveh-Amit H, Berzsenyi S, Vas V, Ye D, Dinnyes A. Tissue resident stem cells: till death do us part. Biogerontology. 2013;**14**(6):573-590. DOI: 10.1007/s10522-013-9469-9

[88] Carlson BM, Faulkner JA. Muscle transplantation between young and old rats: Age of host determines recovery. The American Journal of Physiology. 1989;**256**(6 Pt 1):C1262-C1266. DOI: 10.1152/ajpcell.1989.256.6.C1262

[89] Ryu BY, Orwig KE, Oatley JM, Avarbock MR, Brinster RL. Effects of aging and niche microenvironment on spermatogonial stem cell self-renewal. Stem Cells. 2006;**24**(6):1505-1511. DOI: 10.1634/stemcells.2005-0580

[90] Conboy IM, Conboy MJ, Wagers AJ, Girma ER, Weissman IL, Rando TA. Rejuvenation of aged progenitor cells

by exposure to a young systemic environment. Nature. 2005;**433**(7027):760-764. DOI: 10.1038/nature03260

[91] Villeda SA, Luo J, Mosher KI, Zou B, Britschgi M, Bieri G, et al. The ageing systemic milieu negatively regulates neurogenesis and cognitive function. Nature. 2011;**477**(7362):90-94. DOI: 10.1038/nature10357

[92] Sousa-Victor P, Gutarra S, Garcia-Prat L, Rodriguez-Ubreva J, Ortet L, Ruiz-Bonilla V, et al. Geriatric muscle stem cells switch reversible quiescence into senescence. Nature. 2014;**506**(7488):316-321. DOI: 10.1038/nature13013

[93] Acharyya S, Sharma SM, Cheng AS, Ladner KJ, He W, Kline W, et al. TNF inhibits Notch-1 in skeletal muscle cells by Ezh2 and DNA methylation mediated repression: implications in Duchenne muscular dystrophy. PloS one. 2010;**5**(8):e12479. DOI: 10.1371/journal.pone.0012479

[94] Tasselli L, Zheng W, Chua KF. SIRT6: Novel mechanisms and links to aging and disease. Trends in Endocrinology and Metabolism: TEM. 2017;**28**(3):168-185. DOI: 10.1016/j.tem.2016.10.002

[95] Conboy IM, Conboy MJ, Smythe GM, Rando TA. Notch-mediated restoration of regenerative potential to aged muscle. Science. 2003;**302**(5650):1575-1577. DOI: 10.1126/science.1087573

[96] Gala-Lopez BL, Neiman D, Kin T, O'Gorman D, Pepper AR, Malcolm AJ, et al. Beta cell death by cell-free DNA and outcome after clinical islet transplantation. Transplantation. 2018;**102**(6):978-985. DOI: 10.1097/TP.0000000000002083

[97] Johansson H, Goto M, Dufrane D, Siegbahn A, Elgue G, Gianello P, et al. Low molecular weight dextran sulfate: A strong candidate drug to block IBMIR in clinical islet transplantation. American Journal of Transplantation. 2006;**6**(2):305-312. DOI: 10.1111/j.1600-6143.2005.01186.x

[98] Friedenstein AJ, Chailakhyan RK, Gerasimov UV. Bone marrow osteogenic stem cells: In vitro cultivation and transplantation in diffusion chambers. Cell and Tissue Kinetics. 1987;**20**(3):263-272

[99] Scharstuhl A, Schewe B, Benz K, Gaissmaier C, Buhring HJ, Stoop R. Chondrogenic potential of human adult mesenchymal stem cells is independent of age or osteoarthritis etiology. Stem Cells. 2007;**25**(12):3244-3251. DOI: 10.1634/stemcells.2007-0300

[100] D'Ippolito G, Schiller PC, Ricordi C, Roos BA, Howard GA. Age-related osteogenic potential of mesenchymal stromal stem cells from human vertebral bone marrow. Journal of Bone and Mineral Research. 1999;**14**(7):1115-1122. DOI: 10.1359/jbmr.1999.14.7.1115

[101] Murphy JM, Dixon K, Beck S, Fabian D, Feldman A, Barry F. Reduced chondrogenic and adipogenic activity of mesenchymal stem cells from patients with advanced osteoarthritis. Arthritis and Rheumatism. 2002;**46**(3):704-713. DOI: 10.1002/art.10118

[102] Phinney DG, Kopen G, Righter W, Webster S, Tremain N, Prockop DJ. Donor variation in the growth properties and osteogenic potential of human marrow stromal cells. Journal of Cellular Biochemistry. 1999;**75**(3):424-436

[103] Sun Y, Li W, Lu Z, Chen R, Ling J, Ran Q, et al. Rescuing replication and osteogenesis of aged mesenchymal stem cells by exposure to a young extracellular matrix. FASEB Journal. 2011;**25**(5):1474-1485. DOI: 10.1096/fj.10-161497

[104] NishikawaK,NakashimaT,TakedaS, Isogai M, Hamada M, Kimura A, et al. Maf promotes osteoblast differentiation in mice by mediating the age-related switch in mesenchymal cell differentiation. The Journal of Clinical Investigation. 2010;**120**(10):3455-3465. DOI: 10.1172/JCI42528

[105] Kurtzberg J. Update on umbilical cord blood transplantation. Current Opinion in Pediatrics. 2009;**21**(1):22-29

[106] Cairo MS, Wagner JE. Placental and/or umbilical cord blood: An alternative source of hematopoietic stem cells for transplantation. Blood. 1997;**90**(12):4665-4678

[107] Gluckman E, Broxmeyer HA, Auerbach AD, Friedman HS, Douglas GW, Devergie A, et al. Hematopoietic reconstitution in a patient with Fanconi's anemia by means of umbilical-cord blood from an HLA-identical sibling. The New England Journal of Medicine. 1989;**321**(17):1174-1178

[108] Cairo MS, Rocha V, Gluckman E, Hale G, Wagner J. Alternative allogeneic donor sources for transplantation for childhood diseases: Unrelated cord blood and haploidentical family donors. Biology of Blood and Marrow Transplantation. 2008;**14**(1 Suppl. 1):44-53. DOI: 10.1016/j.bbmt.2007.10.007

[109] Ingram DA, Mead LE, Tanaka H, Meade V, Fenoglio A, Mortell K, et al. Identification of a novel hierarchy of endothelial progenitor cells using human peripheral and umbilical cord blood. Blood. 2004;**104**(9):2752-2760. DOI: 10.1182/blood-2004-04-1396

[110] McGuckin C, Jurga M, Ali H, Strbad M, Forraz N. Culture of embryonic-like stem cells from human umbilical cord blood and onward differentiation to neural cells in vitro. Nature Protocols. 2008;**3**(6):1046-1055. DOI: 10.1038/nprot.2008.69

[111] Kucia M, Halasa M, Wysoczynski M, Baskiewicz-Masiuk M, Moldenhawer S, Zuba-Surma E, et al. Morphological and molecular characterization of novel population of CXCR4+ SSEA-4+ Oct-4+ very small embryonic-like cells purified from human cord blood: preliminary report. Leukemia. 2007;**21**(2):297-303. DOI: 10.1038/sj.leu.2404470

[112] van de Ven C, Collins D, Bradley MB, Morris E, Cairo MS. The potential of umbilical cord blood multipotent stem cells for nonhematopoietic tissue and cell regeneration. Experimental Hematology. 2007;**35**(12):1753-1765. DOI: 10.1016/j.exphem.2007.08.017

[113] Kogler G, Sensken S, Airey JA, Trapp T, Muschen M, Feldhahn N, et al. A new human somatic stem cell from placental cord blood with intrinsic pluripotent differentiation potential. The Journal of Experimental Medicine. 2004;**200**(2):123-135. DOI: 10.1084/jem.20040440

[114] Ratajczak MZ, Ratajczak J, Suszynska M, Miller DM, Kucia M, Shin DM. A novel view of the adult stem cell compartment from the perspective of a quiescent population of very small embryonic-like stem cells. Circulation Research. 2017;**120**(1):166-178. DOI: 10.1161/CIRCRESAHA.116.309362

[115] Danova-Alt R, Heider A, Egger D, Cross M, Alt R. Very small embryonic-like stem cells purified from umbilical cord blood lack stem cell characteristics. PLoS One. 2012;**7**(4):e34899. DOI: 10.1371/journal.pone.0034899

[116] Kluth SM, Buchheiser A, Houben AP, Geyh S, Krenz T, Radke TF, et al. DLK-1 as a marker to distinguish unrestricted somatic stem cells and mesenchymal stromal cells in cord blood. Stem Cells and Development. 2010;**19**(10):1471-1483. DOI: 10.1089/scd.2010.0070

[117] Kogler G, Radke TF, Lefort A, Sensken S, Fischer J, Sorg RV, et al. Cytokine production and hematopoiesis supporting activity of cord blood-derived unrestricted somatic stem cells. Experimental Hematology. 2005;**33**(5):573-583. DOI: 10.1016/j.exphem.2005.01.012

[118] Liedtke S, Buchheiser A, Bosch J, Bosse F, Kruse F, Zhao X, et al. The HOX Code as a "biological fingerprint" to distinguish functionally distinct stem cell populations derived from cord blood. Stem Cell Research. 2010;**5**(1):40-50. DOI: 10.1016/j.scr.2010.03.004

[119] Schira J, Falkenberg H, Hendricks M, Waldera-Lupa DM, Kogler G, Meyer HE, et al. Characterization of regenerative phenotype of unrestricted somatic stem cells (USSC) from human umbilical cord blood (hUCB) by functional secretome analysis. Molecular & Cellular Proteomics. 2015;**14**(10):2630-2643. DOI: 10.1074/mcp.M115.049312

[120] Liao Y, Itoh M, Yang A, Zhu H, Roberts S, Highet AM, et al. Human cord blood-derived unrestricted somatic stem cells promote wound healing and have therapeutic potential for patients with recessive dystrophic epidermolysis bullosa. Cell transplantation. 2014;**23**(3):303-317. DOI: 10.3727/096368913X663569

[121] Lee KS, Cha SH, Kang HW, Song JY, Lee KW, Ko KB, et al. Effects of serial passage on the characteristics and chondrogenic differentiation of canine umbilical cord matrix derived mesenchymal stem cells. Asian-Australasian Journal of Animal Sciences. 2013;**26**(4):588-595. DOI: 10.5713/ajas.2012.12488

[122] Schira J, Gasis M, Estrada V, Hendricks M, Schmitz C, Trapp T, et al. Significant clinical, neuropathological and behavioural recovery from acute spinal cord trauma by transplantation of a well-defined somatic stem cell from human umbilical cord blood. Brain: A Journal of Neurology. 2012;**135**(Pt 2):431-446. DOI: 10.1093/brain/awr222

[123] Ghodsizad A, Ungerer MN, Bordel V, Kallenbach K, Kogler G, Bruckner B, et al. Transplanted human cord blood-derived unrestricted somatic stem cells preserve high-energy reserves at the site of acute myocardial infarction. Cytotherapy. 2011;**13**(8):956-961. DOI: 10.3109/14653249.2011.563290

[124] Ghodsizad A, Fahy BN, Waclawczyk S, Liedtke S, Gonzalez Berjon JM, Barrios R, et al. Portal application of human unrestricted somatic stem cells to support hepatic regeneration after portal embolization and tumor surgery. ASAIO Journal. 2012;**58**(3):255-261. DOI: 10.1097/MAT.0b013e31824cc922

[125] Liao Y, Ivanova L, Zhu H, Plumer T, Hamby C, Mehta B, et al. Cord blood-derived stem cells suppress fibrosis and may prevent malignant progression in recessive dystrophic epidermolysis bullosa. Stem Cells. 2018;**36**(12):1839-1850. DOI: 10.1002/stem.2907

[126] Robin C, Bollerot K, Mendes S, Haak E, Crisan M, Cerisoli F, et al. Human placenta is a potent hematopoietic niche containing hematopoietic stem and progenitor cells throughout development. Cell Stem Cell. 2009;**5**(4):385-395. DOI: 10.1016/j.stem.2009.08.020

[127] Liao Y, Ivanova L, Sivalenka R, Plumer T, Zhu H, Zhang X, et al. Efficacy of human placental-derived stem cells in collagen VII knockout (recessive dystrophic epidermolysis bullosa) animal model. Stem Cells Translational Medicine. 2018;**7**(7):530-542. DOI: 10.1002/sctm.17-0182

[128] Cairo MS, Tarek N, Lee DA, Delaney C. Cellular engineering and

therapy in combination with cord blood allografting in pediatric recipients. Bone Marrow Transplantation. 2016;**51**(1):27-33. DOI: 10.1038/bmt.2015.196

[129] Ballen KK, Gluckman E, Broxmeyer HE. Umbilical cord blood transplantation: The first 25 years and beyond. Blood. 2013;**122**(4):491-498. DOI: 10.1182/blood-2013-02-453175

[130] Escolar ML, Poe MD, Provenzale JM, Richards KC, Allison J, Wood S, et al. Transplantation of umbilical-cord blood in babies with infantile Krabbe's disease. The New England Journal of Medicine. 2005;**352**(20):2069-2081. DOI: 10.1056/NEJMoa042604

[131] Krivit W, Shapiro EG, Peters C, Wagner JE, Cornu G, Kurtzberg J, et al. Hematopoietic stem-cell transplantation in globoid-cell leukodystrophy. The New England Journal of Medicine. 1998;**338**(16):1119-1126. DOI: 10.1056/NEJM199804163381605

[132] Wright MD, Poe MD, DeRenzo A, Haldal S, Escolar ML. Developmental outcomes of cord blood transplantation for Krabbe disease: A 15-year study. Neurology. 2017;**89**(13):1365-1372. DOI: 10.1212/WNL.0000000000004418

[133] Sun J, Allison J, McLaughlin C, Sledge L, Waters-Pick B, Wease S, et al. Differences in quality between privately and publicly banked umbilical cord blood units: A pilot study of autologous cord blood infusion in children with acquired neurologic disorders. Transfusion. 2010;**50**(9):1980-1987. DOI: 10.1111/j.1537-2995.2010.02720.x

[134] Cotten CM, Murtha AP, Goldberg RN, Grotegut CA, Smith PB, Goldstein RF, et al. Feasibility of autologous cord blood cells for infants with hypoxic-ischemic encephalopathy. The Journal of Pediatrics. 2014;**164**(5):973-979 e1. DOI: 10.1016/j.jpeds.2013.11.036

[135] Banerjee MN, Bolli R, Hare JM. Clinical studies of cell therapy in cardiovascular medicine: Recent developments and future directions. Circulation Research. 2018;**123**(2):266-287. DOI: 10.1161/CIRCRESAHA.118.311217

[136] Stamm C, Westphal B, Kleine HD, Petzsch M, Kittner C, Klinge H, et al. Autologous bone-marrow stem-cell transplantation for myocardial regeneration. Lancet. 2003;**361**(9351):45-46. DOI: 10.1016/S0140-6736(03)12110-1

[137] Yerebakan C, Kaminski A, Westphal B, Donndorf P, Glass A, Liebold A, et al. Impact of preoperative left ventricular function and time from infarction on the long-term benefits after intramyocardial CD133(+) bone marrow stem cell transplant. The Journal of Thoracic and Cardiovascular Surgery. 2011;**142**(6):1530-1539.e3. DOI: 10.1016/j.jtcvs.2011.05.002

[138] Stamm C, Nasseri B, Hetzer R. Cardiac stem cells in patients with ischaemic cardiomyopathy. Lancet. 2012;**379**(9819):891. DOI: 10.1016/S0140-6736(12)60385-7

[139] Bolli R, Chugh AR, D'Amario D, Loughran JH, Stoddard MF, Ikram S, et al. Cardiac stem cells in patients with ischaemic cardiomyopathy (SCIPIO): Initial results of a randomised phase 1 trial. Lancet. 2011;**378**(9806):1847-1857. DOI: 10.1016/S0140-6736(11)61590-0

[140] Quevedo HC, Hatzistergos KE, Oskouei BN, Feigenbaum GS, Rodriguez JE, Valdes D, et al. Allogeneic mesenchymal stem cells restore cardiac function in chronic ischemic cardiomyopathy via trilineage differentiating capacity. Proceedings of the National Academy of Sciences of the United States of America. 2009;**106**(33):14022-14027. DOI: 10.1073/pnas.0903201106

[141] Zhang S, Wang D, Estrov Z, Raj S, Willerson JT, Yeh ET. Both cell fusion and transdifferentiation account for the transformation of human peripheral blood CD34-positive cells into cardiomyocytes in vivo. Circulation. 2004;**110**(25):3803-3807. DOI: 10.1161/01.CIR.0000150796.18473.8E

[142] Tang XL, Li Q, Rokosh G, Sanganalmath SK, Chen N, Ou Q, et al. Long-term outcome of administration of c-kit(POS) cardiac progenitor cells after acute myocardial infarction: Transplanted cells do not become cardiomyocytes, but structural and functional improvement and proliferation of endogenous cells persist for at least one year. Circulation Research. 2016;**118**(7):1091-1105. DOI: 10.1161/CIRCRESAHA.115.307647

[143] Chimenti I, Smith RR, Li TS, Gerstenblith G, Messina E, Giacomello A, et al. Relative roles of direct regeneration versus paracrine effects of human cardiosphere-derived cells transplanted into infarcted mice. Circulation Research. 2010;**106**(5):971-980. DOI: 10.1161/CIRCRESAHA.109.210682

[144] Marsh SE, Yeung ST, Torres M, Lau L, Davis JL, Monuki ES, et al. HuCNS-SC human NSCs fail to differentiate, form ectopic clusters, and provide no cognitive benefits in a transgenic model of Alzheimer's disease. Stem Cell Reports. 2017;**8**(2):235-248. DOI: 10.1016/j.stemcr.2016.12.019

[145] Heeschen C, Lehmann R, Honold J, Assmus B, Aicher A, Walter DH, et al. Profoundly reduced neovascularization capacity of bone marrow mononuclear cells derived from patients with chronic ischemic heart disease. Circulation. 2004;**109**(13):1615-1622. DOI: 10.1161/01.CIR.0000124476.32871.E3

[146] Anderson AJ, Piltti KM, Hooshmand MJ, Nishi RA, Cummings BJ. Preclinical efficacy failure of human CNS-derived stem cells for use in the pathway study of cervical spinal cord injury. Stem Cell Reports. 2017;**8**(2):249-263. DOI: 10.1016/j.stemcr.2016.12.018

[147] Guidance for Industry: Preclinical Assessment of Investigational Cellular and Gene Therapy Products; Availability. In: Register F, editor. Rockville, MD: Office of the Federal Register, National Archives and Records Administration; 2013. p. 70307

[148] Daley GQ, Hyun I, Apperley JF, Barker RA, Benvenisty N, Bredenoord AL, et al. Setting global standards for stem cell research and clinical translation: The 2016 ISSCR guidelines. Stem Cell Reports. 2016;**6**(6):787-797. DOI: 10.1016/j.stemcr.2016.05.001

Chapter 6

Stem Cells and Extracellular Vesicles: Biological Regulators of Physiology and Disease

Theo Borgovan, Lorin Crawford, Chibuikem Nwizu and Peter Quesenberry

Abstract

This is an extraordinary time in cell biology with evolving data pushing a reconsideration of the stability of cell systems and the regulatory mechanisms underlying cell phenotypes, especially the functional cell phenotypes. In this chapter, we will explore new insights into stem cell and extracellular vesicle biology with a focus on the role of extracellular vesicles in normal stem cell physiology as well as in various disease states. Extracellular vesicles (EVs) are being recognized as influential mediators of cellular function and potential experimental therapeutic strategies for a number of disorders outlined in this review. An evolving paradigm indicates a dynamic flux of EV populations within these disease states. We conclude our discussion of EV by extending our knowledge of robust EV biology toward disease detection and prognostication. Characterizing the biophysical and functional changes of vesicles amid disease progression or regression enables investigators to merge this information flux with existing deep learning computational and statistical techniques—allowing knowledge to be abstracted from large data sets profiling the biology of EVs within various disease states. Understanding how EV population shifts represent disease regression or progression creates paramount potential for EVs as salient and clinically relevant diagnostic and prognosticating tools.

Keywords: stem cell continuum, extracellular vesicles, deep learning, biomarkers

1. Introduction

In many traditional stem/progenitor models for different tissues the generally accepted models have posited a primitive stem cell giving rise to more differentiated progenitors and finally terminally differentiated end cells, which may or may not retain the capacity for cell division. Perhaps the most intensively studied stem cell system has been that of the hematopoietic stem cell [1–6]. In general, current dogma has it that the long-term repopulating hematopoietic stem cell is a dormant non-cycling cell characterized by a surface phenotype that is negative for conventional differentiation markers (B220, Gr-1, Mac-1, Lyt-2, L3T4 and Ter119) and positive for c-kit, Sca-1 and CD150. It is felt that this cell can be purified by FACS and that in response to various differentiating stimuli it progressively differentiates into different lineage restricted populations, which in general are actively cycling. A large number of studies have extensively characterized its molecular regulation and biologic characteristics [6–11]. We began studying both purified stem cells and unseparated whole marrow stimulated to progress through cell cycle with cytokine exposure and demonstrated that there were cycle related and reversible changes over time in long and short term engraftment, progenitor levels, differentiation into megakaryocytes and granulocytes, homing to marrow, capacity to alter phenotype toward lung cells in response to pulmonary derived extracellular vesicles (EVs), overall gene expression, capacity to take up vesicles and circadian characteristics [12–22]. Passegue and colleagues [23] studied lineage negative Sca-1+ c-kit+ and thy+/− stem cells further separated into G0, G1 and S/G2/M fractions as to long-term engraftment into lethally irradiated mice. They found that all engraftment capacity was in the G0 population. This indicated that our observations might be *in vitro* artifacts. However, we noted that no one had adequately studied unseparated marrow as to the cycle status of long-term repopulating stem cells. We essentially reproduced the Passegue data studying purified stem cells, but when we studied unseparated whole marrow we found that over 50% of the long-term engraftable cells were in the S/G2/M fractions. In order to address the potential problem of cellular cross contamination in the FACS experiments we utilized a thymidine suicide technique in which cycling cells are selectively killed by a 30-minute exposure to high specific activity tritiated thymidine. Studies here showed that anywhere between 70 and 100% of the long-term engrafting stem cells were in S phase at the time of the incubation. Further work using *in vivo* BrdU indicated that the dormant purified stem cells (Lin-C-kit+Sc-1+) rapidly progressed through the cell cycle such that up to 85% of them showed the BrdU label by 48 hours of *in vivo* BrdU exposure. These data showed that a large number of HSC in the mouse are actively proliferating and thus always changing phenotype. When lineage positive and lineage negative marrow cells were assessed for engraftment and cycle status, it was found that a large number of marrow stem cells were in both fractions and were cycling [24]. In further work, we have shown that different lineage positive cells are rich in cycling stem cells, but intriguingly when double sorted that lineage positive cells no longer showed HSC characteristics but there was a separate population negative for the particular marker with enriched and cycling HSC. This data has led to our current hypothesis that hematopoietic stem cells exist on a cycle related continuum and that these cells while maintaining critical stem cell markers show cycle related fluctuation in differentiation markers [22]. We feel that we may be defining the calculus of hematopoietic stem cells with time and cycle related phenotype changes being the derivatives and with the final overall picture the integral.

Whether this model of small incremental cellular changes over time applies to other stem cell systems will clearly be the object of much future work.

The stability of cell types and systems is also up for grabs. Early in the study of hematopoietic stem cells Till et al. [25] showed that the first stem cell assay, the colony forming unit spleen, showed marked heterogeneity but the assay was generally reliable. They compared the CFU-s system to radioactive isotopes; the individual decay rates were totally heterogeneous, but the overall half-lives were reproducible and quite exact. These data would indicate the potential importance of evaluating the total population of stem cells alongside the purified variety. The variable and shifting phenotypes of the stem cell with cell cycle transit has to be considered in the context of extracellular vesicle modulation of cell phenotype.

1.1 Extracellular vesicles

Tiny lipid membrane enclosed particles are released from essentially all cell types in the mammalian body [26, 27]. These entities were first found to come from red blood cells and platelets and felt to be essential cellular waste products [28, 29]. Subsequent work has characterized and subdivided these entities to size, density and morphology. Eventually two basic types of vesicles were defined by differential ultracentrifugation; exosomes and microvesicles. Exosomes derived from multivesicular bodies were from 30 to 100 nm in diameter while microvesicles derived from membrane blebbing were from 100 to 1000 nm in diameter. Other vesicular entities were also defined including apoptotic bodies. In general, there was much overlap between exosomes and microvesicular bodies and a meeting of investigators decided it might be best to simply term these as extracellular vesicles (EV) and then define their source and the conditions under which they were isolated. Vesicles were eventually isolated from virtually all bodily fluids and cells [26, 27]. Recent focus has been on the capacity of EV to restore injured tissue and treat disease. Initial work showed that vesicles could transfer protein and RNA while modifying the phenotype of cells and reversing disease in animal models.

Ratajczak et al. showed that embryonic stem cell derived microvesicles could reprogram hematopoietic progenitors by horizontal transfer of mRNA and protein delivery [30]. This was followed by work by Aliotta et al. [31–33] and Valadi et al. [34] showing RNA transfer and phenotype change in different experimental models. Further work has indicated that cellular phenotype change may be mediated by transfer of transcriptional activators possibly miRNA [35]. Vesicles from different sources are different but do contain some features of their originating cells. Vesicles contain protein, mRNA, miRNA, lipids and variably DNA, thus they represent complex bio machines with a tremendous range of potential phenotype altering messages.

1.2 Hematopoietic stem/progenitor cells

Vesicles have been found to have a variety of effects on both normal and diseased or injured tissues. These effects may be negative or positive depending upon the specifics of the experimental models under consideration. In many instances there appears to be a yin/yang nature to vesicle effects. Studies with murine marrow cells have illustrated the complexity of vesicle marrow cell interactions [26, 27]. Early studies showed that ES derived vesicles could improve proliferative status of lin-Sca-1+ marrow stem cells [30] and work in our laboratory has shown that lung derived EVs could induce expression of surfactant A, B, C and D, Clara cell protein and aquaporin in normal murine marrow cells [31–33]. Studies indicated that for a genomic change to occur the vesicle had to enter the target marrow cells [34] and that initially both mRNA and a transcriptional regulator were transferred to target cells but that the transferred mRNA was degraded and long-term expression of surfactants B and C (those tested in these studies) derived from the target cells and represented a stable epigenetic event [35]. The functional effects of vesicles on marrow mRNA expression depended upon the cell cycle status of the target marrow cells and the condition of the originator lung cells, in this case either irradiated or not [21]. The results showed that Lin-Sca-1 murine marrow cells showed peak pulmonary epithelial cell-specific mRNA expression in cell cycle phase G0/G1 when the vesicles were derived from irradiated lung tissue while the peak was in late G1/early S phase when the vesicles were derived from nonirradiated lung. Vesicles were present in all types of differentiated marrow cells. Vesicles demonstrated a capacity to reverse radiation damage to marrow and gastrointestinal tissues of mice with the most impressive effect being on long-term engrafting stem cells [36]. The vesicles were shown to increase proliferation, decrease apoptosis and reverse double-strand DNA breaks.

2. EVs and implications on selected disease states

In the following section, we discuss the role of EVs in various disease states, as well as their role in disease detection, progression, and treatment.

2.1 Pulmonary hypertension

The yin/yang vesicle effect is clearly illustrated by studies on murine models of pulmonary hypertension.

There are two major models of murine pulmonary hypertension; the monocrotaline treated mouse [37] and the Sugen/hypoxia treated mouse [38]. These may represent different forms of pulmonary hypertension but results with different vesicle populations have been similar in both models. Work with the monocrotaline murine model has shown that vesicles in the serum or from the lungs of mice with monocrotaline induced pulmonary hypertension will induce pulmonary hypertension when injected into normal mice [39]. Further work has indicated that marrow from these mice with PH will induce PH in normal irradiated mice [40]. It appears that vesicles from damaged lung tissue, probably damaged endothelium, travel to marrow and induce an endothelial to hematopoietic transition (EHT) with production of "toxic" endothelial progenitors which travel back to the lung, differentiate into pro inflammatory macrophages and induce vascular remodeling resulting in pulmonary hypertension [41]. Marrow derived mesenchymal stem cell (MCS) derived extracellular vesicles were shown to either prevent or reverse pulmonary hypertension in both rodent models [39]. As the endothelial progenitors are quite radiosensitive low dose was tested as a potential therapy for pulmonary hypertension. One hundred cGy whole body irradiation both prevents and reverses pulmonary hypertension in these models. This is the second potential therapy. The EHT is regulated to a large extent by the transcriptional factor Runx-1. A Runx-1 inhibitor, RO5-3335 has been investigated in leukemia. Here we have shown that the Runx-1 inhibitor Ro5-3335 blocks the EHT and reverses pulmonary hypertension in the rodent models [41]. Thus, three potential therapies have evolved from extracellular vesicle research.

2.2 Vesicles in renal disorders

Dr. Giovanni Camussi and colleagues have carried out a series of ground breaking studies on MSC-vesicle effects in murine models of kidney injury. They demonstrated that MSC-vesicles could stimulate proliferation and diminish apoptosis of injured kidney cells [42, 43]. Human mRNAs were transferred and translated into proteins in renal epithelial tubular cells of kidney injured mice. They studied cisplatin treated mice with acute kidney damage. Here they found a dose related correction of injury and felt the therapeutic action was related to the antiapoptotic effect of the MSC-vesicles. They also investigated a ischemia-reperfusion model of kidney injury and showed that the injury could be prevented by a single infusion of MSC-EV [44]. These workers also demonstrated that the active vesicle population were the smaller exosomes as opposed to microvesicles [45].

2.3 Lung cancer

Being one of the most common, and increasingly deadly cancers in the world, lung cancer lends itself to early screening mechanisms, as well as the potential clinical value EVs may hold in diagnosing and treating neoplastic disease [46, 51, 52, 54].

Exosomal nucleic acid (such as microRNAs) released from neoplastic lung cancer cells play a vital role in cancer's ability to evade immune response. These cancer-derived exosomes have been shown to have a critical impact on disease progression via their ability to modulate gene expression post-transcriptionally [47, 50]. Lung cancer derived exosomes are laden with, and shuttle a vast array of immune suppressive cargo that stymie the function of immune cells. Interestingly the protein and nucleic acids carried in these tumor-derived exosomes is similar to those of the parent cell it was derived from, allowing for an effective mode of non-contact-dependent cellular manipulation which has wide reaching implication on cancer immune evasion and metastasis.

Vesicles can also have direct actions on target cells. Tumor-associated antigens are also loaded into, and perhaps found bound to the surface of many of these nano-lipid carriers, which can then go on to directly modulate immune mediators' cellular function [47, 71, 73, 74, 86].

EVs shed from lung cancer have also had various implications on the tumor microenvironment and phenotype, a phenomenon observed across numerous cancer types including leukemia [67]. Vascular endothelial growth factor (VEGF) has often been studied as a potential drug target to quench the growth of localized and distant lung cancer. Certain monoclonal antibodies that target VEGF are used to inhibit the formation of new vasculature often initiated by growing cancer cells, which in essence starves a growing tumor from oxygen and other vital nutrients [49]. Work by Azmi has shown how a selective population of EVs allows sensitive lung cancer to escape these treatments. Tumor cells threatened by an increasingly more hypoxic microenvironment secrete a very select population of adapted EVs which can directly stimulate the formation of new blood vessels as well entire organelles, including mitochondria, allowing for a more efficient biochemical use and energy production within an oxygen depleted microenvironment [50]. Other works have confirmed this showing that STAT3-regulated exosomal miR-21 enhances the level of VEGF.

Hypoxia and other cellular stressors can also drive numerous cellular adaptations in lung and other cancers. Hypoxia, acidosis, an immune response initiation such as with endotoxin promotes tumor cells to secret more oncogenic EV—these cancer-derived exosomes have direct toles in mediating metastasis; perhaps being implicated in early cellular dysregulation in order to establish a pre-metastatic niche for future metastatic cells [53].

There is direct evidence for the involvement of exosomes from highly metastatic cancer cells in educating stromal cells and altering the cancer microenvironment. In addition, much of the stromal microenvironment that is exposed to cancer undergoes mesenchymal transition (EMT), allowing for the genesis of a more aggressive phenotype via an EV-mediated process [55]. Rahman et al. found that exosomes derived from patients with lung cancer induced vimentin expression, and subsequent EMT in normal lung epithelial cells [56].

Lung cancer derived exosomes promote cancer survival via a myriad of other mechanisms, including fibroblast growth to enhance desmoplastic stromal response which has been shown to enhance tumor growth and block drug delivery in lung, breast and pancreatic cancer models. In addition, tumor cell derived EVs can sequester and carry bioactive Fas ligand (FasL) which has a role in inducing immune cell death, thus dampening the T cell immune response and progressing metastasis in lung cancer [57].

Prior work has explored the effect of EVs from lung and bone marrow sources, and demonstrated that once at the effector cell, EVs impart cellular effects by several purported mechanisms including: (i) direct binding and activation of cell surface receptors by proteins and lipid ligands, or (ii) fusion and uptake (phagocytosis/endocytosis) of vesicle contents into the recipient cells. Effector molecules (e.g., mRNA), non-coding regulatory RNAs (e.g., microRNAs or miRNAs), proteins, and transcription factors can all be delivered, each having short- and long-term implications on effector cell phenotype and function [58, 59]. As discussed, various other studies have also highlighted the ability of EVs to directly transfer relatively larger molecules such as cellular receptors, major histocompatibility complex (MHC) molecules, antigens, as well as entire organelles, some containing fully intact mitochondria, lysosomes, Golgi and intermediate filaments [60].

2.4 Breast

EV studies relating to both breast and prostate cancer highlight many of the salient principles observed in lung cancer studies, and also exhibit the promising roles that EV play in evolving chemoresistance. As we will come to see across a variety of cancer models and disease states, EV function carries great plurality-exhibiting multiple, and often times contradictory effects depending on their cellular origin and physiological state [48].

In breast cancer, while healthy mammary epithelial cells within the breast stroma secrete EVs that prevent the release of breast cancer derived EVs, the EVs shed by the disease cells promote the opposite, imparting an immense impact on chemoresistance. Cancer derived EVs are known to shuttle pro-oncogenic proteins and nucleic acids from diseases cells to surrounding healthy stroma and connective tissue [61]. Zhou et al. reported that breast cancer secreted exosomes are enriched in particular RNA species, such as miR-105, which destroys the vascular endothelial barrier, allowing cancer to enter the circulation and spread [62]. Studies employing fluorescently labeled miRNA-loaded EVs showed that tamoxifen resistant breast cancer cells *in vitro* can carry multiple miRNA profiles. EVs packed with fluorescently-tagged miR-221/222 can also shuttle their cargo to sensitive cells of the same type, thereby transferring resistance RNAs which effectively reduced gene expression of P27 and estrogen receptor-α (ERα) in target cell. The loss of p27 has been linked to drug resistance, as it is able to take a cell that is arrested in its cell cycle and stimulate its reentry back into active cycling [63]. However, as discussed, healthy stromal cells counteract the effects of oncogenic vesicles. This competition between "good" and "bad" vesicles is a fine balance; a yin/yang that loses equilibrium as cancer overwhelms healthy stoma. When stromal cells are outcompeted and significantly influenced by oncogenic EV signaling, the now altered stroma in turn activates STAT1 and NOTCH3 signaling in breast cancer cells, promoting cancer initiating cell populations responsible for drug resistance and nascent tumor formation [64]. This is a common theme in EV-mediated cancer progression which we will see is universal across numerous solid and hematological cancers.

2.5 Prostate cancer

Human bone marrow mesenchymal stem cell (MSC) derived EVs are involved in the modulation of cell signaling, cellular differentiation, and proliferation—and this is seen across multiple disease paradigms. These regenerative EVs have been shown to reverse the malignant phenotype in prostate and colorectal cancer, recovering function in a murine model of AKI, as well as mitigating radiation damage to marrow [36]. In models of prostate cancers, the reversal of taxane resistance and tumorigenic phenotype in a human prostate carcinoma cell line (as well as human explants) can be accomplished by treatment with healthy MSC-derived EVs [61].

Other populations of "therapeutic" EVs (outside of the bone marrow) have also been isolated and applied: EVs isolated from normal prostate cells acquired via patient biopsy reverse the resistance of malignant prostate cells to various drugs. On the contrary we have shown that EVs derived from cancerous cells can drive cancer progression and enhance resistance to certain chemotherapies, which again highlights the specificity, plurality, the yin/yang of EV functionality [61]. Panagopoulos et al. confirmed much of this work, showing that vesicles from both *in vivo* prostate cancer cell and explant cultured prostate cancer cells can induce cellular changes that produce a neoplastic phenotype in normal prostate cell lines [65]. These results were also reproduced using vesicles from patients with other malignancies, namely prostate, and lung [59].

2.6 Neural-derived EVs in traumatic brain injury

Our group has developed a unique biomarker system focusing on EVs isolated from the saliva of patients who have experienced mild traumatic brain injury (mTBI) [66]. Rather than conventional human serum isolation, this has been a novel protocol allowing for the easily accessible collection of saliva laden with EVs that have freely trafficked from injured brain parenchyma into the saliva—allowing for a representative sample which captures the shift of various EV populations and cargo following brain trauma. EVs are membrane bound, and thus are not subject to the same degradation that conventional serum biomarkers face, making them ideal biomarker candidates. Salivary EVs in particular can be isolated based on tissue specificity and have well established roles in the detection of numerous other disease states, including oral squamous cell carcinoma [79]. Bolstering their utility as a unique biomarker, upon analysis and characterization of patients that had sustained mTBI it became apparent that EVs isolated from saliva had numerous neural markers on them, confirming their origin from brain parenchyma [66]. Following analysis of the expression of Alzheimer disease (AD) genes in patients who had suffered mTBI vs. healthy controls, multiple important AD specific genes were significantly upregulated in patients that had suffered mTBI when compared to healthy controls; allowing for the identification of mTBI-specific genetic profiles derived from neural derived EVs. The potential characterization of early mTBI biomarker genes, including (but not limited to) CTSD, CDC2 and casein kinase (CSNK1A1) is being explored [66]. Longitudinal analysis of these patients coupled with further analysis of the identified surrogate markers allows for possible prognostication of mTBI patients in regard to severity of post-TBI concussion symptoms, chronicity of symptomology and potential recovery. This is all made possible by the ubiquitous and specific nature of EVs.

2.7 Hematologic malignancies

EVs secreted by blood borne hematologic cancer have modulating affects impacting a variety of cancer hallmarks. EVs have a direct effect on phenotypic and genotypic changes, highlighting the central role of EVs in the progressions and reversal of hematologic malignancies.

2.7.1 Impact on various leukemias

Pathways involved in angiogenesis have been shown to modulate cancer progressions and chemotherapeutic evasion in multiple models [61]. Vesicles shed by chronic lymphocytic and myelogenous leukemia (CLL, CML) transmit cargo containing a myriad of cancer-inducing factors, such as rapamycin/p70S6K/ hypoxia-inducible factor-1α axis. Similar to lung cancer-derived EVS, these CML-derived vesicles have been shown to bolster the survival of CML B-cells via the establishment and proliferation of vascular endothelial growth factor within the forming leukemic bone marrow stromal cells [67, 69]. In multiple myeloma (MM) models, bone marrow stromal cell-derived exosomes, mediate cellular communication by transferring mRNAs, miRNAs, and proteins important in proliferation, survival, and chemoresistance [77]. Experiments utilizing *in vivo* mouse K562 CML cells showed that neogenic angiogenesis can be induced by immortalized myelogenous leukemia cell line K562 exosomes, as well as neogenic changes in human umbilical endothelial cells [69].

Other hematologic malignancies show similar cancer induction potential. CML-derived EVs given to rat models can induce CML-like characteristics via the transfer of their oncogenic cargo [67]. Bone marrow stromal cells respond to this influence by producing interleukin (IL)-8 (mRNA and protein), a potent pro-angiogenic factor that modulates both *in vitro* and *in vivo* the leukemia cell malignant phenotype [67]. In our own established acute myeloid leukemia (AML) model, we explored the potential of human bone marrow MSC-derived EVs as a direct adjunct therapy for AML. Our studies indicated that the killing potential of cytarabine, at even relatively low doses, is potentiated by the addition of healthy MSC-derived EVs. We believe EVs can also alter a cancer cell's sensitivity to chemotherapy via EV guided horizontal information transfer. This has implications directly on the cell itself but likely also impacts the surrounding stroma in order to further promote oncogenic growth and drug resistance of leukemia cells [68]. In models of MM when marrow MSC-derived exosomes are cultured with cancerous MM cells, there is a significant increase in multiple anti-apoptotic pathways which promoted MM cell viability. These exosomes, derived from stromal cells within a microenvironment amidst developing active cancer, were also able to induce drug resistance to the proteasome inhibitor bortezomib via activation of several survival relevant pathways, including c-Jun N-terminal kinase, p38, p53, and Akt [97].

2.7.2 Impact on chemo-resistance

Cancer derived EVs play a central role in facilitating the escape of cell death, by cancer cell. Proteins such as BCL-2, MCL-1, BCL-X, and BAX as well as other cell death-related proteins were shown to be more concentrated in the EV of apoptosis-resistant primary AML blasts than EVs from more sensitive AML cells [70]. Via confocal-microscopy-based colocalization studies, the direct transfer of EVs from resistant to sensitive cells has been observed. Leukemia derived EV harbor multiple bioactive lipids, proteins and miRNAS important in chemoresistance. Ibrutinib is a drug used clinically to combat leukemia. Analysis of plasma samples collected from CLL patients showed exosomes bearing unique micro-RNA prolife, including miR-29 family, miR-150, miR-155, and miR-223, showed a different exosome profile from what is seen when disease is suppressed with ibrutinib treatment—perhaps, indicating the potential pathophysiology by which cancerous EVs impart resistance, as well as creating a potential for biomarker identification [75]. EVs packed with miR-221/222, from tamoxifen resistant MCF-7 breast cancer cells, can shuttle their cargo to sensitive cells of the same type, thereby transferring resistance.

2.7.3 Impact on the cancer microenvironment

Healthy bone marrow stroma likely functions to maintain and protect healthy bone marrow stroma from nascent cancer. At first detection of threat, the bone marrow microenvironment and residing cells, such as MSCs, combat early cancer—much of this is likely EV mediated. Our established leukemic cell model has established this hypothesis, indicating that MSC-EVs impart a robust anti-proliferative and pro-apoptotic effect on leukemic cells *in vitro*. We also have preliminary data using EVs toward clinically relevant endpoints, and have showed they serve as a synergistic adjunct to conventional AML therapies, such as cytarabine.

As discussed, the bone marrow stroma can be recapitulated by active cancer via multiple EV-dependent mechanisms. Leukemic models have shown the net effect of EV-modulation translated to a phenotypic change of the bone marrow stromal cells toward a more inflammatory signature that resembles the phenotype of cancer-associated fibroblasts (CAFs) [72]. CAFs show enhanced proliferation, migration, and secretion of inflammatory cytokines, all contributing to a tumor-supportive niche [72]. As a result, stromal cells exposed to a leukemic EVs are not killed but "reprogrammed" to be pro-oncogenic and support tumor growth. As discussed, EV populations change depending on disease state [48]. In the case of CLL, as leukemic cancer cells progress varying EV populations establish control within the microenvironment. CLL-derived EVs rapidly deliver their biologic cargo to the surrounding stromal cells, and promoting CAF phenotypes with enhanced proliferative and migratory properties [72]. CLL models have shown that CAF-derived factors may also have an immunogenic effect on the T and myeloid cells, altering their phenotypes into immunosuppressive and tumor-promoting Th2/M2-like cells, respectively. These modifications lead to defective T-cell and myeloid cell immune responses and an inflammatory milieu characteristic of CLL promotion [75].

Leukemic EVs impart genotypic and phenotypic effects on all components of the leukemic microenvironment. The bony endosteal compartment of the bone marrow niche, composed of osteocytes/osteoblasts/osteoclasts, is reprogrammed by AML—derived EVs toward inflammatory myelofibrotic cells. These cells support leukemic growth and support BM fibrosis, a well-established risk factor for leukemia [75]. Metastasis is crucial to cancer survival. Leukemia derived EVs have also been shown to disturb the architecture of multiple tight junction proteins in cells of the basement membrane, allowing cancer to detach, mobilize, and metastasize beyond in situ disease. Leukemic EVs can also bolster angiogenesis [75]. *In vitro* studies, first reported by Umezu et al., clearly showed leukemic cell to endothelial cell communication via exosomal miRNAs by fluorophore signaling localization, allowing for the creation of new blood vessels to feed cancer growing in newly seeded microenvironmental niches [76]. Microenvironment stromal cells have been shown to directly take up cargo from EVs fluorescently labeled with GFP. As we've seen in lung and breast cancer cells, Boelens et al. showed that this cross talk is reciprocal and that when stromal cells are influenced by oncogenic EV signaling the stromal cells themselves in turn activate STAT1 and NOTCH3 signaling in developing cancer cells. This cell signaling in turn leads to cell populations responsible for drug resistance and nascent tumor formation [64, 78]. Schepers et al. have shown that AML cells (likely via an EV directed mechanism) cause numerous chromosomal anomalies and genetic mutations within the surrounding stroma, thereby altering the biology of the stem cell continuum away from normal hematopoiesis, and toward transforming bone marrow stem cells toward immature progenitors that will subsequently develop into leukemic blasts or altered cancer-stem cells capable of supporting a pro-leukemic environment [75]. The CAF phenotype promoted by tumor-derived EVs, has, itself, secondary effects on endothelial cells, increasing angiogenesis. The sum and synergy of all of these EV-directed microenvironment modulations means the leukemia-modified stroma favors leukemic blast proliferation while stymieing normal hematopoiesis [69, 76].

3. Machine learning

In this section, we discuss machine learning (ML) as an emerging scientific field of sophisticated algorithms that aid in the understanding of how nonlinear interactions between molecular features contribute to disease etiology. Here, we give relevant background on how machine learning is used in biology, provide a formal and probabilistic specification of the hierarchical architectures implemented by common ML methods (Bayesian deep neural networks), and demonstrate their power via real data applications. Indeed, a myriad of well-established algorithms can be surveyed in detail, but our main goal is to develop a more conceptual pipeline on how to use machine learning techniques on individualized biological problems.

In the context of our own research interests, we have found vesicle biology to be amenable to ML because of (i) the ability to observe millions of vesicles during a single study and (ii) the nonlinear nature of downstream vesicle effects. As we will show, large sample sizes and the presence of variable interactions are often leveraged by ML algorithms to provide high predictive accuracies. We hypothesize that these performance gains will lend to a more complete picture of how vesicle behavior impacts the overall cellular environment.

3.1 Background and significance

Machine learning is often described as a subarea of artificial intelligence that seeks to recognize subtle patterns found within data. It has been noted that the field has roots dating back to early work done by Arthur Samuel in 1959 [80]. However, despite this long history, only recent technological advances over the past two and a half decades have considerably revived interest in ML. With increases in both data collection and computational power, the applications for machine learning algorithms have become vast and integral parts of our everyday lives (e.g., facial recognition, spam detection, etc.)

One explanation for the utility of ML approaches is that they are able to model complex structures in data and leverage the detailed information to accurately predict or classify unobserved outcomes. Unique to these algorithms is their ability to adaptively update themselves (learning) through repeated exposure to new observations (a process formally known as "training") [81, 82]. Intuitively, an algorithm should achieve a higher predictive accuracy after training on larger data sets: the more possibilities an algorithm is exposed to, the better said algorithm becomes at correctly identifying similar complex patterns in heterogeneous populations [82, 85]. This represents a common tenet about ML theory: the more data the better. However, just having data is not always enough. A second tenet is related to the strength of signal between the observed data and the scientific question of interest. The greater the signal-to-noise ratio, the more amenable the task is to ML methodology. In practice, there exists a general relationship between tenets 1 and 2: the more data one has, the less robust the signal-to-noise ratio must be to achieve an acceptable prediction/classification; conversely, a high signal-to-noise ratio will compensate for less data. Note that this is obviously not a strict relationship, as many have demonstrated ML algorithms to perform well on noisy data sets with few observations.

With its growing popularity in the biological literature, the formal connection between machine learning and more traditional statistical sciences cannot be overlooked. Indeed, many current approaches in ML are motivated by prediction; however, there are opportunities to pair these tools with fundamental probabilistic concepts to improve power for inference-based tasks as well. This is particularly relevant for biological problems where it is also important to understand the processes that are contributing to better predictions. To this end, recent works have used (interpretable) ML algorithms for live risk stratification in cancer patients [6], novel biomarker identification in liquid biopsies [87], hypoxemia prevention during surgery [88], point-of-care diagnosis of lymphoma [89], as well as many other uses in genetics and genomics [82, 83].

3.2 Probabilistic formulation

With an increasing literature on both statistical and machine learning methods, it can be difficult to decide which algorithm to use **Figure 1** provides a general approach for determining the proper choice [90]. In this section, however, we will focus on detailing an increasingly popular machine learning method known as a neural network (NN). Although NNs excel at classification tasks (see **Figure 1**), many recent works have focused on applying neural networks to a wider range of applications [83, 91, 92].

For simplicity, we will consider an arbitrary data analysis problem. Let y be an n-dimensional response/outcome vector for n individuals. Assume that for each individual, we measure p features and tabulate their collection via an $n \times p$ design matrix X. Statistically, these features are variables that we believe will help accurately predict the outcome. In the case of our research on vesicle biology, features may be biophysical (i.e., vesical diameter and volume), genomic (i.e., sequence data), proteomic, or lipidomic measurements. Following previous work, we may specify a (Bayesian) NN by assuming some hierarchical architecture to "learn" the predicted response for each observation in the data [96].

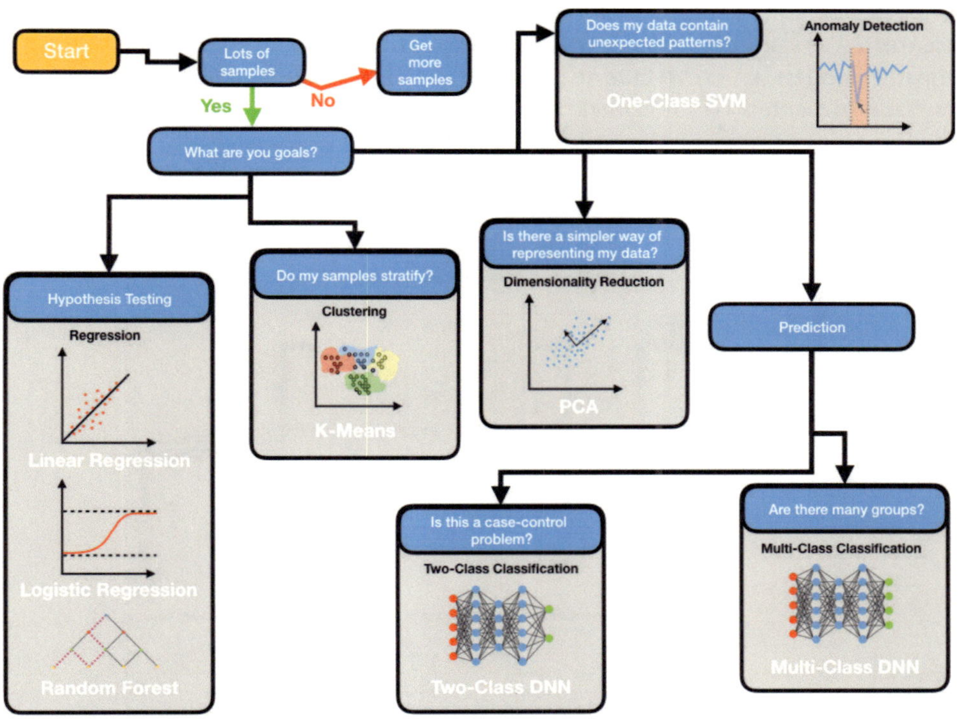

Figure 1.
Basic "decision tree" schematic for deciding between different statistical and machine learning methods. Here, approaches are grouped by their designed purpose for applications.

$$\hat{y} = \sigma[f] \tag{1}$$

$$f = H(\theta)w + b \tag{2}$$

$$w \sim \pi \tag{3}$$

These sets of equations reformulate a general NN as a probabilistic hierarchical statistical model. In Eq. (1), $\hat{y}$ is an n-dimensional vector of predicted values, f is an n-dimensional vector of continuous unbounded values that need to be estimated, and $\sigma[\cdot]$ is a link function that relates f to the mean of the (assumed) distribution of y. Note that the link function can be flexibly changed depending on the goals of the research. For example, in the case of regression problems with continuous outcomes, the link function is set to the identity; while for classification-based applications with binary data, we may use a sigmoid function that which transforms the systematic part of the model to be between 0 and 1. If one is faces with a multiclass problem, then $\sigma[\cdot]$ can be redefined as a softmax function.

In Eq. (2), we use $H[\theta]$ to denote an $n \times k$ matrix of activations from the penultimate layer (which are fixed given a set of inputs and point estimates θ from previous layers), w is a k-dimensional vector of weights at the output layer that is assumed to follow some prior distribution π (see Eq. (3)), and b is an n-dimensional vector of biases that is produced during the training phase.

Under this formulation, notice that we may divide arbitrary Neural Networks into three components (see the middle panel in **Figure 2**): (i) an input layer of the p features in the design matrix X (red nodes), (ii) a set of hidden layers where parameters are deterministically computed based off of a set series of activations and point estimates (blue nodes), and (iii) a penultimate layer where the weights are treated as random variables (green nodes). This structure is also highly generalizable: hidden layers can take on any form, provided that the additional structure can be represented via some linear combination of activations, weights, and biases.

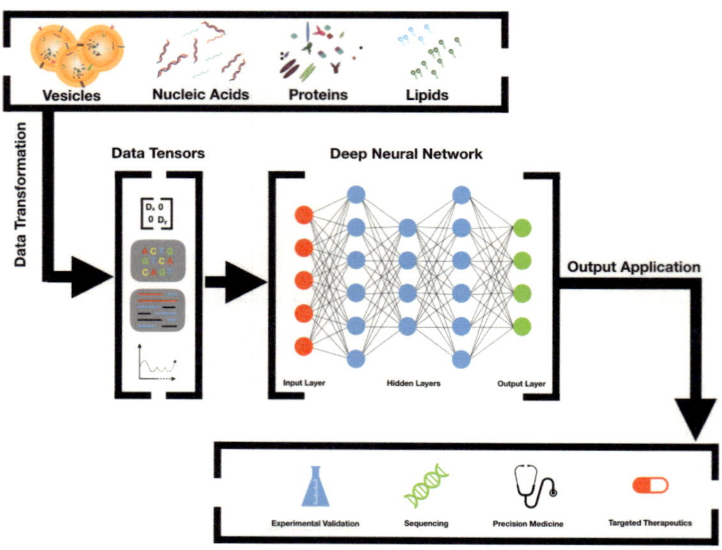

Figure 2.
Our general work flow for when using neural networks for prediction purposes in biological datasets. Here, we show how diverse feature types can be transformed/quantified and used for various applications.

3.3 Real data applications

We now demonstrate how machine learning and, more specifically, neural networks can be adopted to positively impact data analysis. Our group looks to characterize the vesicle phenotype of patients at various stages of treatment in various leukemias, such as AML. Here, we utilize a common NN architecture known as a Multilayer Perceptron [84] where we first train the algorithm on patients with known disease statuses (i.e., $y_i = 1$ if the ith patient has cancer) and then test its ability to accurately classify a set of undiagnosed individuals. We define accuracy here as simply the percentage of correctly classified samples in a testing dataset. For each validation run, a Receiver Operating Characteristic (ROC) curve is drawn and the area under the curve (AUC) is calculated. The AUC is a standard performance metric for classification problems in statistics and may be interpreted as an assessment of how effective an algorithm is at discriminating between two classes (i.e., a healthy versus a disease phenotype) [93]. Higher AUC values (on a scale from 0 to 100%) indicate better model performance. An overall summary of our workflow may be found in **Figure 2** where we illustrate how different biological features are quantified and few through a NN to make predictions.

We first trained the algorithm on data collected from a NanoSight Tracking instrument, the NS5000. This allowed us to collect a wide selection of vesicle features including size, area, volume, diffusion coefficients, and total vesicles secreted. This data was collected from two cell type populations: (i) a primary hMSC cell line and (ii) a Kasumi AML cell line. We did this in order to first assess the validity of the idea that there is a discernable difference between vesicles derived from "normal" hMSC and vesicles from the cancerous Kasumi cell line. Within the training set, we were able to classify vesicles with relatively high accuracy: the mean AUC (plus or minus standard deviation) after 10-fold cross validation was 90.16 ± 9.26%. This translated into a high accuracy in the testing population with a mean AUC (after 10-fold cross validation) of 95.97 ± 5.38%. We next tested the algorithm on real patient samples, achieving perfect accuracy in reliably characterizing and classifying healthy tissue. We believe the reason for the high accuracy is due to the primary hMSC cell line accurately representing the vesicular phenotype of normal, healthy bone marrow. There is still some work to be done in accurately classifying malignant samples. We believe that the heterogeneity of the leukemic vesicle phenotype cannot trivially be captured through cell line data [94, 95].

To address this heterogeneity problem, we then elected to train and test our machine learning algorithm solely on patient samples—in hopes of increasing the predictive performance. We fed the model 35 samples from patients with various hematologic conditions. We tested and trained the model on these 35 samples and were able to achieve a mean training accuracy of 93.76 ± 4.77% and an out-of-sample AUC of 97.33 ± 3.46%. The high testing performance suggests that the algorithm is capable of accurate classification and serves as a general proof-of-concept of the potential utility of machine learning in this space. Here, this technology has the power to identify complex, heterogeneous patterns that distinguish the normal healthy vesicle phenotypes from leukemic vesicle phenotypes.

4. Conclusion

EVs are a ubiquitous and dynamic population of cell-specific information. Functionally, they act as a class of membrane-bound cellular communication particles that contain bioactive molecules. By exerting their effects through RNA, proteins, lipids and variably DNA, EVs implement various downstream phenotypic and genotypic effects across multiple disease states. By enabling contact-free cell to cell communication, EVs can modulate normal physiological homeostasis. Moreover, research has shown that certain subpopulations of EVs are responsible for initiating and maintaining certain pathological states. This is the "Yin and Yang" of vesicle biology which posits that EV populations harbor a functional endpoint specific to cell type and disease state. There are far reaching implication in utilizing EVs toward clinical endpoints focused on disease identification, progression, modulation, and ultimately cures.

In order for EVs to be effectively utilized in the identification, prognostication, modulation, and curing of disease, more work needs to be done with regards to understanding the nonlinear effects of EVs on target cells. Enter the need for novel and sophisticated statistical modeling techniques. The steadily increasing size of "omic" data sets, along with significant improvements in computational power and machinery, has caused resurging interest in machine learning. Consequently, this revival has led to algorithmic improvements in both predictive accuracy and precision. With the large amount of information that EVs provide, there is a unique opportunity to gain new knowledge from applying ML techniques to current problems focused on better understanding complex EV biology.

Our lab has already begun utilizing ML algorithms in the characterization of diseased subpopulations of EVs. Currently, most active areas of research in vesicle biology focus on characterizing EVs via isolation methods. Alternatively, we propose to analyze entire populations of EVs jointly, for we believe a wholistic view better captures the true nature and variability of a patient's disease process. Thus, our work is novel in this respect: we use predictive algorithms to identify subtle patterns within a given vesicle population. Here, we analyze how particular subpopulations interact and detail how these interactions influence the underlying disease process. Furthermore, we hypothesize that by monitoring a patient's entire population of EVs throughout the course of treatment, we can better predict the efficacy of the treatment. Preliminary results have yielded positive results with respect to categorizing diseased and healthy EV populations. Work now must be done to further characterize these subpopulations within the context of specific diseases, such as AML and other blood neoplasms. EV biology presents another avenue of utility for the field of machine learning. Concatenating large sets of EV information within interpretable ML algorithmic frameworks can lead us closer to the use of EVs as a predictive and useful clinical marker. Overall, the future is bright for both the fields of EV biology and ML.

DOI: http://dx.doi.org/10.5772/intechopen.86845

Author details

Theo Borgovan[1*], Lorin Crawford[2], Chibuikem Nwizu[1] and Peter Quesenberry[1]

1 Division of Hematology and Oncology, Rhode Island Hospital, The Warren Alpert Medical School of Brown University, Providence, RI, USA

2 Center for Computational Molecular Biology, Brown University School of Public Health, USA

*Address all correspondence to: tb2182@columbia.edu

References

[1] Rossi L, Challen GA, Sirin O, Lin KK, Goodell MA. Hematopoietic stem cell characterization and isolation. Methods in Molecular Biology. 2011;**750**:47-59

[2] Spangrude GJ, Heimfeld S, Weissman IL. Purification and characterization of mouse hematopoietic stem cells. Science. 1988;**241**:58-62

[3] Morrison SJ, Wandycz AM, Hemmati HD, Wright DE, Weissman IL. Identification of lineage of multipotent hematopoietic progenitors. Development. 1997;**124**:1929-1939

[4] Adolfsson J, Mansson R, Buza-Vidas N, Hultquist A, Liuba K, Jensen CT, et al. Identification of Flt3+ lympho-myeloid stem cells lacking erythromegakaryocytic potential: A revised road map for adult blood lineage commitment. Cell. 2005;**121**:295-306

[5] Forsberg EC, Serwold T, Kogan S, Weissman IL, Passegue E. New evidence supporting megakaryocyte-erythrocyte potential of flk2/flt3+ multipotent hematopoietic progenitors. Cell. 2006;**126**:415-426

[6] Yimez OH, Kiel MJ, Morrison SJ. Slam family markers are conserved among hematopoietic stem cells from old and reconstituted mice and markedly increase their purity. Blood. 2006;**107**:924-930

[7] Varnum-Finney B, Brashem-Stein C, Bernstein ID. Combined effects of Notch signaling and cytokines induce a multiple log increase in precursors with lymphoid and myeloid reconstituting ability. Blood. 2003;**101**:1784-1789

[8] Poulos MG, Guo P, Kofler NM, Pinho S, Gutkin MC, Tikhonova A, et al. Endothelial Jagged-1 is necessary for homeostatic and regenerative hematopoiesis. Cell Reports. 2013;**4**:1022-1034

[9] Maillard I, Koch U, Dumortier A, Shestova O, Xu L, Sai H, et al. Canonical notch signaling is dispensable for the maintenance of adult hematopoietic stem cells. Cell Stem Cell. 2008;**2**:356-366

[10] Reya T, Duncan AW, Ailles L, Domen J, Scherer DC, Willert K, et al. A role for Wnt signalling in self-renewal of haematopoietic stem cells. Nature. 2003;**423**:409-414

[11] Scheller M, Huelsken J, Rosenbauer F, Taketo MM, Birchmeier W, Tenen DG, et al. Hematopoietic stem cell and multilineage defects generated by constitutive β-catenin activation. Nature Immunology. 2006;**7**:1037-1047

[12] Peters SO, Kittler EL, Ramshaw HS, Quesenberry PJ. *Ex vivo* expansion of murine marrow cells with interleukin-3 [IL-3], IL-6, IL-11, and stem cell factor leads to impaired engraftment in irradiated hosts. Blood. 1996;**87**:30-37

[13] Habibian HK, Peters SO, Hsieh CC, Wuu J, Vergilis K, Grimaldi CI, et al. The fluctuating phenotype of the lympho-hematopoietic stem cell with cell cycle transit. The Journal of Experimental Medicine. 1998;**188**:393-398

[14] Egan PC, Liang OD, Goldberg LR, et al. Low dose 100 cGy irradiation as a potential therapy for pulmonary hypertension. Journal of Cellular Physiology. 2019:1-6

[15] Margolin DA, Myers T, Zhang X, Bertoni DM, Reuter BA, Obokhare I, et al. The critical roles of tumor-initiating cells and the lymph node stromal microenvironment in human colorectal cancer extranodal metastasis using a unique humanized orthotopic mouse model. The FASEB Journal. 2015;**29**(8):3571-3581

[16] Reddy GP, Tiarks CY, Pang L, Wuu J, Hsieh CC, Quesenberry PJ. Cell cycle analysis and synchronization of pluripotent hematopoietic progenitor stem cells. Blood. 1997;**90**:2293-2299

[17] Colvin GA, Dooner MS, Dooner GJ, Sanchez-Guijo FM, Demers DA, Abedi M, et al. Stem cell continuum: Directed differentiation hotspots. Experimental Hematology. 2007;**35**:96-107

[18] Cerny J, Dooner M, McAuliffe C, Habibian H, Stencil K, Berrios V. Homing of purified murine lympho-hematopoietic stem cells: A cytokine-induced defect. Journal of Hematotherapy & Stem Cell Research. 2002;**11**:913-922

[19] Becker PS, Nilsson SK, Li Z, Berrios VM, Dooner MS, Cooper CL, et al. Adhesion receptor expression by hematopoietic cell lines and murine progenitors: Modulation by cytokines and cell cycle status. Experimental Hematology. 1999;**27**:533-541

[20] Colvin GA, Lambert JF, Moore BE, Carlson JE, Dooner MS, Abedi M, et al. Intrinsic hematopoietic stem cell/progenitor plasticity: Inversions. Journal of Cellular Physiology. 2004;**199**:20-31

[21] Aliotta JM, Lee D, Puente N, Faradyan S, Sears EH, Amaral A, et al. Progenitor/stem cell fate determination: Interactive dynamics of cell cycle and microvesicles. Stem Cells and Development. 2012;**21**:1627-1638

[22] Quesenberry PJ, Colvin GA, Abedi M, Dooner G, Dooner M, Aliotta J, et al. The stem cell continuum. Annals of the New York Academy of Sciences. 2005;**1044**:228-235

[23] Passegue E, Wagers AJ, Giuriato S, Anderson WC, Weissman IL. Global analysis of proliferation and cell cycle gene expression in the regulation of hematopoietic stem and progenitor cell fates. The Journal of Experimental Medicine. 2005;**202**:1599-1611

[24] Goldberg LR, Dooner MS, Johnson K, Papa E, Pereira M, Del Tatto M, et al. The murine long-term multi-lineage renewal marrow stem cell is a cycling cell. Leukemia. 2014;**28**:813-822

[25] Till JE, McCulloch EA, Siminovitch L. A stochastic model of stem cell proliferation, based on the growth of spleen colony-forming cells. Proceedings of the National Academy of Sciences of the United States of America. 1964;**51**:29-36

[26] Quesenberry PJ, Aliotta J, Deregibus MC, Camussi G. Role of extracellular RNA-carrying vesicles in cell differentiation and reprogramming. Stem Cell Research & Therapy. 2015;**6**:153. DOI: 10.1186/s13287-015-0150-x. Review

[27] Quesenberry PJ, Goldberg LR, Aliotta JM, Dooner MS, Pereira MG, Wen S, et al. Cellular phenotype and extracellular vesicles: Basic and clinical considerations. Stem Cells and Development. 2014;**23**(13):1429-1436. DOI: 10.1089/scd.2013.0594. Epub: 1 April 2014

[28] Heijnen HF, Schiel AE, Fijnheer R, Geuze HJ, Sixma JJ. Activated platelets release two types of membrane vesicles: Microvesicles by surface shedding and exosomes derived from exocytosis of multivesicular bodies and alpha-granules. Blood. 1999;**94**:3791-3799

[29] Johnstone RM, Adam M, Hammond JR, Orr L, Turbide C. Vesicle formation during reticulocyte maturation. Association of plasma membrane activities with released vesicles [exosomes]. The Journal of Biological Chemistry. 1987;**262**:9412-9420

[30] Ratajczak J, Miekus K, Kucia M, Zhang J, Reca R, Dvorak P, et al. Embryonic stem cell-derived microvesicles reprogram hematopoietic progenitors: Evidence for horizontal transfer of mRNA and protein delivery.

Leukemia. 2006;**20**:847-856. DOI: 10.1038/sj.leu.2404132

[31] Aliotta JM, Keaney P, Passero M, Dooner MS, Pimentel J, Greer D, et al. Bone marrow production of lung cells: The impact of G-CSF, cardiotoxin, graded doses of irradiation, and subpopulation phenotype. Experimental Hematology. 2006;**34**:230-241

[32] Aliotta JM, Sanchez-Guijo FM, Dooner GJ, Johnson KW, Dooner MS, Greer KA, et al. Alteration of marrow cell gene expression, protein production, and engraftment into lung by lung-derived microvesicles: A novel mechanism for phenotype modulation. Stem Cells. 2007;**25**:2245-2256

[33] Aliotta JM, Pereira M, Johnson KW, de Paz N, Dooner MS, Puente N, et al. Microvesicle entry into marrow cells mediates tissue-specific changes in mRNA by direct delivery of mRNA and induction of transcription. Experimental Hematology. 2010;**38**:233-245

[34] Valadi H, Ekstrom K, Bossios A, Sjostrand M, Lee JJ, Lotvall JO. Exosome-mediated transfer of mRNAs and microRNAs is a novel mechanism of genetic exchange between cells. Nature Cell Biology. 2007;**9**:654-659. DOI: 10.1038/ncb1596

[35] Aliotta JM, Pereira M, Sears EH, Dooner MS, Wen S, Goldberg LR, et al. Lung-derived exosome uptake into and epigenetic modulation of marrow progenitor/stem and differentiated cells. Journal of the Extracellular Vesicles. 2015;**4**:26166

[36] Wen S, Dooner M, Cheng Y, Papa E, Del Tatto M, Pereira M, et al. Mesenchymal stromal cell-derived extracellular vesicles rescue radiation damage to murine marrow hematopoietic cells. Leukemia. 2016;**30**(11):2221-2231. DOI: 10.1038/leu.2016.107. Epub 6 May 2016

[37] Nogueira-Ferreira R, Vitorino R, Ferreira R, Henriques-Coelho T. Exploring the monocrotaline animal model for the study of pulmonary arterial hypertension: A network approach. Pulmonary Pharmacology & Therapeutics. 2015;**35**:8-16

[38] Vitali SH, Hansmann G, Rose C, Fernandez-Gonzalez A, Scheid A, Mitsialis SA, et al. The Sugen 5416/hypoxia mouse model of pulmonary hypertension revisited: Long-term follow-up. Pulmonary Circulation. 2014;**4**(4):619-629

[39] Aliotta JM, Pereira M, Wen S, Dooner MS, Del Tatto M, Papa E, et al. Exosomes induce and reverse monocrotaline-induced pulmonary hypertension in mice. Cardiovascular Research. 2016;**110**(3):319-330

[40] Aliotta JM, Pereira M, Wen S, Dooner MS, Del Tatto M, Papa E, et al. Bone marrow endothelial progenitor cells are the cellular mediators of pulmonary hypertension in the murine monocrotaline injury model. Stem Cells Translational Medicine. 2017;**6**(7):1595-1606. DOI: 10.1002/sctm.16-0386. Epub 5 May 2017

[41] Liang OD, So EY, Egan PC, Goldberg LR, Aliotta JM, Wu KQ, et al. Endothelial to haematopoietic transition contributes to pulmonary arterial hypertension. Cardiovascular Research. 2017;**113**(13):1560-1573

[42] Gaetti S, Bruno S, Deregibus MC, Sordi A, Cantaluppi V, Tetta C, et al. Microvesicles derived from human adult mesenchymal stem cells protect against ischemia-reperfusion-induced acute and chronic kidney injury. Nephrology, Dialysis, Transplantation. 2011;**5**:1474-1483

[43] Bruno S, Grange C, Deregibus MC, Calogero RA, Saviozzi S, Collino F, et al. Mesenchymal stem

cell-derived microvesicles protect against acute tubular injury. Journal of the American Society of Nephrology. 2009;**5**:1053-1067

[44] Gatti S, Bruno S, Deregibus MC, Sordi A, Cantaluppi V, Tetta C, et al. Microvesicles derived from human adult mesenchymal stem cells protect against ischaemia-reperfusion-induced acute and chronic kidney injury. Nephrology, Dialysis, Transplantation. 2011;**26**:1474

[45] SWen S, Quesenberry P, Camussi G. Renal regenerative potential of different extracellular vesicle populations derived from bone marrow mesenchymal stromal cells. Tissue Engineering. Part A. 2017;**23**(21-22):1262-1273. DOI: 10.1089/ten.TEA.2017.0069. Epub: 13 June 2017

[46] Cheng G. Circulating miRNAs: Roles in cancer diagnosis, prognosis and therapy. Advanced Drug Delivery Reviews. 2015;**81**:75-93

[47] Zhang X, Yuan X, Shi H, Wu L, Qian H, Xu W. Exosomes in cancer: Small particle, big player. Journal of Hematology & Oncology. 2015;**8**:83

[48] Dooner MS, Stewart C, Deng Y, et al. Daily rhythms influence the ability of lung-derived extracellular vesicles to modulate bone marrow cell phenotype. PLoS One. 2018;**13**(11):e0207444. DOI: 10.1371/journal.pone.0207444. Published 26 November 2018

[49] Liu Y, Luo F, Wang B, Li H, Xu Y, Liu X, et al. STAT3-regulated exosomal miR-21 promotes angiogenesis and is involved in neoplastic processes of transformed human bronchial epithelial cells. Cancer Letters. 2016;**370**(1):125-135

[50] Azmi AS, Bao B, Sarkar FH. Exosomes in cancer development, metastasis and drug resistance: A comprehensive review. Cancer Metastasis Reviews. 2013;**32**(3-4):623-642

[51] Lindoso RS, Collino F, Bruno S, Araujo DS, Sant'Anna JF, Tetta C, et al. Extracellular vesicles released from mesenchymal stromal cells modulate miRNA in renal tubular cells and inhibit ATP depletion injury. Stem Cells and Development. 2014;**23**:1809-1819

[52] Quesenberry PJ, Morley A, Stohlman F, et al. Effect of endotoxin on granulopoiesis and the *in vitro* colony-forming cell. Blood. 1973;**15**:18-26

[53] King HW, Michael MZ, Gleadle JM. Hypoxic enhancement of exosome release by breast cancer cells. BMC Cancer. 2012;**12**:421

[54] Suetsugu A, Honma K, Saji S, Moriwaki H, Ochiya T, Hoffman RM. Imaging exosome transfer from breast cancer cells to stroma at metastatic sites in orthotopic nude-mouse models. Advanced Drug Delivery Reviews. 2013;**65**(3):383-390

[55] Blackwell RH, Foreman KE, Gupta GN. The role of cancer-derived exosomes in tumorigenicity and epithelial-to-mesenchymal transition. Cancers [Basel]. 2017;**9**(8):1102-1111

[56] Rahman MA, Barger JF, Lovat F, Gao M, Otterson GA, Nana-Sinkam P. Lung cancer exosomes as drivers of epithelial mesenchymal transition. Oncotarget. 2016;**7**(34):54852-54866

[57] Becker A, Thakur BK, Weiss JM, Kim HS, Peinado H, Lyden D. Extracellular vesicles in cancer: Cell-to-cell mediators of metastasis. Cancer Cell. 2016;**30**(6):836-848

[58] Mathivanan S, Ji H, Simpson RJ. Exosomes: Extracellular organelles important in intercellular communication. Journal of Proteomics. 2010;**73**(10):1907-1920

[59] Aliotta JM, Pereira M, Sears EH, et al. Lung-derived exosome uptake into and epigenetic modulation of marrow progenitor/stem and differentiated cells. Journal of the Extracell Vesicles. 2015;**4**:26166. DOI: 10.3402/jev.v4.26166. Published 16 September 2015

[60] Johnson SM, Dempsey C, Parker C, Mironov A, Bradley H, Saha V. Acute lymphoblastic leukaemia cells produce large extracellular vesicles containing organelles and an active cytoskeleton. Journal of the Extracell Vesicles. 2018;**3**:1809-1819

[61] Wang JQ, DeChalus A, Chatterjee DN, et al. Extracellular vesicle-mediated reversal of paclitaxel resistance in prostate cancer. Critical Reviews in Oncogenesis. 2015;**20**(5-6):407-417

[62] Zhou W, Fong MY, Min Y, Somlo G, Liu L, Palomares MR, et al. Cancer-secreted miR-105 destroys vascular endothelial barriers to promote metastasis. Cancer Cell. 2014;**25**(4):501-515

[63] Zhu Z, Zhang D, Lee H, et al. Macrophage-derived apoptotic bodies pro-mote the proliferation of the recipient cells via shuttling microRNA-221/222. Journal of Leukocyte Biology. 2017;**101**(6):1349-1359

[64] Whiteside TL. Immune modulation of T-cell and NK (natural killer) cell activities by TEXs (tumour-derived exosomes). Biochemical Society Transactions. 2013;**41**(1):245-251

[65] Panagopoulos K, Cross-Knorr S, Dillard C, et al. Reversal of chemosensitivity and induction of cell malignancy of a non-malignant prostate cancer cell line upon extracellular vesicle exposure. Molecular Cancer. 2013;**12**(1):118

[66] Cheng Y, Pereira M, Raukar N, et al. Potential biomarkers to detect traumatic brain injury by the profiling of salivary extracellular vesicles. Journal of Cellular Physiology. 2018;**234**:14377-14388

[67] Corrado C, Saieva L, Raimondo S, Santoro A, De Leo G, Alessandro R. Chronic myelogenous leukaemia exosomes modulate bone marrow microenvironment through activation of epidermal growth factor receptor. Journal of Cellular and Molecular Medicine. 2016;**20**(10):1829-1839

[68] Fei F, Joo EJ, Tarighat SS, et al. B-cell precursor acute lymphoblastic leukemia and stromal cells communicate through Galectin-3. Oncotarget. 2015;**6**(13):11378-11394

[69] Mineo M, Garfield SH, Taverna S, et al. Exosomes released by K562 chronic myeloid leukemia cells promote angiogenesis in a Src-dependent fashion. Angiogenesis. 2011;**15**(1):33-45

[70] Viola S, Traer E, Huan J, et al. Alterations in acute myeloid leukaemia bone marrow stromal cell exosome content coincide with gains in tyrosine kinase inhibitor resistance. British Journal of Haematology. 2015;**172**(6):983-986

[71] Pomatto MAC, Gai C, Bussolati B, Camussi G. Extracellular vesicles in renal pathophysiology. Frontiers in Molecular Biosciences. 2017;**4**:37. DOI: 10.3389/fmolb.2017.00037. Published 7 June 2017

[72] Pando A, Reagan JL, Quesenberry P, Fast LD. Extracellular vesicles in leukemia. Leukemia Research. 2018;**64**:52-60

[73] Aliotta JM, Pereira M, Wen S, et al. Exosomes induce and reverse mono-crotaline-induced pulmonary hypertension in mice. Cardiovascular Research. 2016;**110**(3):319-330

[74] Crompot E, Van Damme M, Pieters K, et al. Extracellular vesicles of bone marrow stromal cells rescue chronic lymphocytic leukemia B cells from apoptosis, enhance their migration and induce gene expression modifications. Haematologica. 2017;**102**(9):1594-1604

[75] Schepers K, Campbell TB, Passegué E. Normal and leukemic stem cell niches: Insights and therapeutic opportunities. Cell Stem Cell. 2015;**16**(3):254-267

[76] Zhou J, Wang S, Sun K, Chng WJ. The emerging roles of exosomes in leukemogeneis. Oncotarget. 2016;**7**(31):50698-50707

[77] Umezu T, Imanishi S, Azuma K, et al. Replenishing exosomes from older bone marrow stromal cells with miR-340 inhibits myeloma-related angio-genesis. Blood Advances. 2017;**1**(13):812-823

[78] Boelens MC, Wu TJ, Nabet BY, Xu B, Qiu Y, Yoon T, et al. Exosome transfer from stromal to breast cancer cells regulates therapy resistance pathways. Cell. 2014;**159**:499-513

[79] Gai C, Camussi F, Broccoletti R, Gambino A, Cabras M, Molinaro L, et al. Salivary extracellular vesicle-associated miRNAs as potential biomarkers in oral squamous cell carcinoma. BMC Cancer. 2018;**18**(1):439. DOI: 10.1186/s12885-018-4364-z

[80] Bzdok D. Classical statistics and statistical learning in imaging neuroscience. Frontiers in Neuroscience. 2017;**11**:23

[81] Ghahramani Z. Probabilistic machine learning and artificial intelligence. Nature. 2015;**521**(7553):452-459

[82] Libbrecht MW, Noble WS. Machine learning applications in genetics and genomics. Nature Reviews Genetics. 2015;**16**(6):321-332

[83] Zou J et al. A primer on deep learning in genomics. Nature Genetics. 2019;**51**(1):12-18

[84] Boulesteix AL, Schmid M. Machine learning versus statistical modeling. Biometrical Journal. 2014;**56**(4):588-593

[85] Bzdok D, Altman N, Krzywinski M. Statistics versus machine learning. Nature Methods. 2018;**15**(4):233-234

[86] Manak MS et al. Live-cell phenotypic-biomarker microfluidic assay for the risk stratification of cancer patients via machine learning. Nature Biomedical Engineering. 2018;**2**(10):761-772

[87] Heitzer E et al. Current and future perspectives ofliquid biopsies in genomics-driven oncology. Nature Reviews. Genetics. 2019;**20**(2):71-88

[88] Lundberg SM et al. Explainable machine-learning predictions for the prevention of hypoxaemia during surgery. Nature Biomedical Engineering. 2018;**2**(10):749-760

[89] Im H et al. Design and clinical validation of a point-of-care device for the diagnosis of lymphoma via contrast-enhanced microholography and machine learning. Nature Biomedical Engineering. 2018;**2**(9):666-674

[90] Noble WS. What is a support vector machine? Nature Biotechnology. 2006;**24**(12):1565-1567

[91] LeCun Y, Bengio Y, Hinton G. Deep learning. Nature. 2015;**521**(7553):436-444

[92] Cao CS et al. Deep learning and its applications in biomedicine. Genomics, Proteomics & Bioinformatics. 2018;**16**(1):17-32

[93] Bradley AP. The use of the area under the roc curve in the evaluation of

machine learning algorithms. Pattern Recognition. 1997;**30**(7):1145-1159

[94] Burdall SE et al. Breast cancer cell lines: Friend or foe? Breast Cancer Research. 2003;**5**(2):89-95

[95] Gazdar AF, Gao BN, Minna JD. Lung cancer cell lines: Useless artifacts or invaluable tools for medical science? Lung Cancer. 2010;**68**(3):309-318

[96] Ish-Horowicz J, Udwin D, Flaxman SR, Filippi SL, Crawford L. Interpreting deep neural networks through variable importance. arXiv, 2019. 1901.09839

[97] Shin DH, Chun Y-S, Lee DS, Huang LE, Park J-W. Bortezomib inhibits tumor adaptation to hypoxia by stimulating the FIH-mediated repression of hypoxia-inducible factor-1. Blood. 2008;**111**(6):3131-3136. DOI: 10.1182/blood-2007-11-120576